**TEACHER'S PET PUBLICATIONS**

LITPLAN TEACHER PACK
for
Speak
based on the book by
Laurie Halse Anderson

Written by
Christina Stone

© 2007 Teacher's Pet Publications
All Rights Reserved

Copyright Teacher's Pet Publications 2007

Only the student materials in this unit plan (such as worksheets, study questions, and tests) may be reproduced multiple times for use in the purchaser's classroom.

For any additional copyright questions, contact Teacher's Pet Publications.

www.tpet.com

TABLE OF CONTENTS - *Speak*

Introduction	5
Unit Objectives	7
Reading Assignment Sheet	8
Unit Outline	9
Study Questions (Short Answer)	13
Quiz/Study Questions (Multiple Choice)	21
Pre-reading Vocabulary Worksheets	37
Lesson One (Introductory Lesson)	51
Writing Assignment 1	63
Writing Evaluation Form	64
Oral Reading Evaluation Form	67
Writing Assignment 2	69
Nonfiction Assignment Sheet	73
Writing Assignment 3	77
Extra Writing Assignments/Discussion ?s	80
Vocabulary Review Activities	93
Unit Review Activities	94
Unit Tests	99
Unit Resource Materials	139
Vocabulary Resource Materials	161

A FEW NOTES ABOUT THE AUTHOR
LAURIE HALSE ANDERSON

Laurie Halse Anderson was born in upstate New York. Her maiden name, Halse, is often mispronounced; it rhymes with waltz.

Laurie was turned on to writing in the second grade during a haiku lesson. She read all the time, enjoying historical fiction, science fiction, and fantasy more than any other genres. She also had an interest in foreign cultures and languages. In high school, Laurie moved out of her parents' house and lived on a pig farm in Denmark where she was an exchange student.

After her high school graduation, Laurie worked for minimum wage in a clothing store, where she quickly discovered she needed to go college. She attended Onondaga Community College for two years, working on a dairy farm milking cows during that time. She later transferred to Georgetown University and graduated with a degree in Languages and Linguistics. Laurie got married after graduating from college; she and her husband then had two daughters.

Laurie never thought of writing as a career, but later became a freelance reporter. Though she was sent several rejection letters, she eventually published some nonfiction and children's storybooks.

In 1999, Laurie published *Speak*, which won several awards. She then published *Fever 1793*, several books that are part of the *Wild at Heart* series, *Prom*, *Catalyst*, which takes place at Merryweather High and references Melinda in her sophomore year at school, and *Twisted*, which was published in 2007.

Recently, Laurie and her husband divorced; she is now remarried and part of a great extended family.

INTRODUCTION

This LitPlan has been designed to develop students' reading, writing, thinking, and language skills through exercises and activities related to *Speak*. It includes twenty lessons, supported by extra resource materials.

The **introductory lesson** introduces students to a symbol used throughout the novel–a tree. Following the introductory activity, students are given a transition to explain how the activity relates to the book they are about to read. Following the transition, students are given the materials they will be using during the unit. At the end of the lesson, students begin the pre-reading work for the first reading assignment.

The **reading assignments** are approximately thirty pages each; some are a little shorter while others are a little longer. Students have approximately 15 minutes of pre-reading work to do prior to each reading assignment. This pre-reading work involves reviewing the study questions for the assignment and doing some vocabulary work for 10 vocabulary words they will encounter in their reading.

The **study guide questions** are fact-based questions; students can find the answers to these questions right in the text. These questions come in two formats: short answer or multiple choice. The best use of these materials is probably to use the short answer version of the questions as study guides for students (since answers will be more complete), and to use the multiple choice version for occasional quizzes.

The **vocabulary work** is intended to enrich students' vocabularies as well as to aid in the students' understanding of the book. Prior to each reading assignment, students will complete a two-part worksheet for approximately 10 vocabulary words in the upcoming reading assignment. Part I focuses on students' use of general knowledge and contextual clues by giving the sentence in which the word appears in the text. Students are then to write down what they think the words mean based on the words' usage. Part II nails down the definitions of the words by giving students dictionary definitions of the words and having students match the words to the correct definitions based on the words' contextual usage. Students should then have an understanding of the words when they meet them in the text.

After each reading assignment, students will go back and formulate answers for the study guide questions. Discussion of these questions serves as a **review** of the most important events and ideas presented in the reading assignments.

After students complete reading the work, there is a **vocabulary review** lesson which pulls together all of the fragmented vocabulary lists for the reading assignments and gives students a review of all of the words they have studied.

Following the vocabulary review, a lesson is devoted to the **extra discussion questions/writing assignments**. These questions focus on interpretation, critical analysis and personal response, employing a variety of thinking skills and adding to the students' understanding of the novel.

There is a **group theme project** in this unit. This project requires students to conduct research on the issue of rape, in hopes of raising awareness about it.

There are three **writing assignments** in this unit, each with the purpose of informing, persuading, or expressing personal opinions. The first writing assignment mimics an essay Melinda and her classmates do in their English class. The second writing asks students to persuade the school board to increase rape and sexual harassment education in schools, and the third gives students several roles in which to write from.

There is a **nonfiction reading assignment**. Students must read nonfiction articles, books, etc. to gather information about their themes in our world today.

The **review lesson** pulls together all of the aspects of the unit. The teacher is given four or five choices of activities or games to use which all serve the same basic function of reviewing all of the information presented in the unit.

The **unit test** comes in two formats: multiple choice or short answer. As a convenience, two different tests for each format have been included. There is also an advanced short answer unit test for advanced students.

There are additional **support materials** included with this unit. The **Unit Resource Materials** section includes suggestions for an in-class library, crossword and word search puzzles related to the novel, and extra worksheets. There is a list of **bulletin board ideas** which gives the teacher suggestions for bulletin boards to go along with this unit. In addition, there is a list of **extra class activities** the teacher could choose from to enhance the unit or as a substitution for an exercise the teacher might feel is inappropriate for his/her class. **Answer keys** are located directly after the **reproducible student materials** throughout the unit. The **Vocabulary Resource Materials** section includes similar worksheets and games to reinforce the vocabulary words.

The **level** of this unit can be varied depending upon the criteria on which the individual assignments are graded, the teacher's expectations of his/her students in class discussions, and the formats chosen for the study guides, quizzes and test. If teachers have other ideas/activities they wish to use, they can usually easily be inserted prior to the review lesson.

The student materials may be reproduced for use in the teacher's classroom without infringement of copyrights. No other portion of this unit may be reproduced without the written consent of Teacher's Pet Publications, Inc.

UNIT OBJECTIVES - *Speak*

1. Through reading Laurie Halse Anderson's novel *Speak*, students will look at the two major symbols—trees and silence— and discuss their roles in the novel.

2. Students will demonstrate their understanding of the text on four levels: factual, interpretive, critical, and personal.

3. Students will make connections with the material in the text and apply the lessons learned to their lives.

4. Students will be given the opportunity to practice reading aloud and silently to improve their skills in each area.

5. Students will answer questions to demonstrate their knowledge and understanding of the main events and characters in *Speak* as they relate to the author's theme development.

6. Students will enrich their vocabularies and improve their understanding of the novel through the vocabulary lessons prepared for use in conjunction with the novel.

7. The writing assignments in this unit are geared to several purposes:
    a. To have students demonstrate their abilities to inform, to persuade, or to express their own personal ideas
        Note: Students will demonstrate the ability to write effectively to inform by developing and organizing facts to convey information. Students will demonstrate the ability to write effectively to persuade by selecting and organizing relevant information, establishing an argumentative purpose, and by designing an appropriate strategy for an identified audience. Students will demonstrate the ability to write effectively to express personal ideas by selecting a form and its appropriate elements.
    b. To check the students' reading comprehension
    c. To make students think about the ideas presented by the novel
    d. To encourage logical thinking
    e. To provide an opportunity to practice good grammar and improve students' use of the English language.

8. Students will read aloud, report, and participate in large and small group discussions to improve their public speaking and personal interaction skills.

READING ASSIGNMENT SHEET - *Speak*

Date Assigned	Pages Assigned	Completion Date
	Assignment #1 Welcome to Merryweather High-Dinner Theater	
	Assignment #2 Blue Roses-Third Marking Period	
	Assignment #3 Death of the Wombat-My Report Card	
	Assignment #4 Exterminators-Oprah, Sally Jessy, Jerry, and Me	
	Assignment #5 Real Spring-Final Cut	

UNIT OUTLINE - *Speak*

1 Introduction Activity Project Assignment PVR1	2 Study Quest. 1 Review Vocab 1 KWL Chart	3 Review Key Literature Terms Review Works Cited and Parenthetical Documentation PVR2	4 Study Quest. 2 Review Vocab 2 Writing Assignment #1	5 Symbolism Art Project PVR3
6 Study Quest. 3 Vocab Review 3 Oral Reading Evaluation PVR4	7 Speaker	8 Writing Assignment #2	9 Study Quest. 4 Vocab Review 4 Figurative Language Activity	10 Nonfiction Assignment PVR5
11 Study Quest. 5 Vocab Review 5 Writing on the Bathroom Stall	12 Writing Assignment #3	13 Movie	14 Movie Cont. Discussion on similarities and differences between book and movie	15 Extra Discussion Quest.
16 Extra Discussion Quest. Cont. Letter to the Author	17 Project Presentations Unit Evaluation	18 Vocabulary Review	19 Unit Review	20 Unit Test

Key: P = Preview Study Questions   V = Vocabulary Work   R = Read

STUDY GUIDE QUESTIONS

SHORT ANSWER STUDY GUIDE QUESTIONS – *Speak*

Assignment #1
Welcome to Merryweather High-Dinner Theater
1. What happens in the cafeteria that makes Melinda run out of the lunchroom?
2. How is Melinda's attitude in her art class different from her attitude towards her other classes?
3. Describe the globe Mr. Freeman uses in his class, and explain how he uses it.
4. Describe how the students at Merryweather High School treat Melinda. Give at least three examples from the text to support your answer.
5. How does Melinda's family communicate?
6. Describe Melinda's new hiding place at school.
7. What did Melinda do that has caused everyone in school to hate her?

Assignment #2
Blue Roses-Third Marking Period
1. What book does Melinda read on Halloween instead of going trick-or treating?
2. Describe Melinda's lips.
3. The school has decided to change the school mascot four times. List all three of the previous mascots that have been used and explain why the school felt they had to change each one.
4. What author has the school district banned from the school library?
5. Explain why David Petrakis walks out of Mr. Neck's class. Give enough details in your answer to fully explain why he is upset.
6. Describe the art project Melinda makes with the turkey bones.
7. Why does David Petrakis video his social studies class?
8. Why is Melinda so overwhelmed by the Christmas gift her parents gave her?
9. Who is the "IT" Melinda refers to?

Assignment #3
Death of the Wombat-My Report Card
1. What animal does Melinda compare herself to whenever she sees Andy?
2. What reason does Heather give Melinda for no longer sitting with her at lunch?
3. Who gives Melinda a Valentine card on Valentine's Day?
4. Who is the one person that Melinda speaks to?
5. Melinda takes inspiration from Picasso in her tree drawing. What unique artistic method does she take from him and apply to her tree?
6. What is Melinda's reward for going to all her classes for an entire week?
7. Describe Melinda's conflict over the party David Petrakis is throwing at his house to celebrate the basketball victory.
8. What was Melinda's initial reaction to Andy when he approached her at the party?

Speak Study Questions page 2

Assignment #4
Exterminators-Oprah, Sally Jessy, Jerry, and Me
1. Why does the PTA want to change the hornet mascot to something else?
2. How does Ivy offer Melinda help?
3. Why does Melinda's dad always take a taxi to the airport?
4. Using details from the text, describe Melinda's conflict over whether or not to warn Rachel about Andy.
5. How does Melinda warn Rachel about Andy?
6. What does Melinda do to get put in in-school suspension?
7. What event prompts Melinda to go home, crawl in her closet, bury her face in her clothes, and scream at the top of her lungs?

Assignment #5
Real Spring-Final Cut
1. What happens at school that ruins Melinda's shirt?
2. Describe how students decorate the bathroom stall doors.
3. What has happened to the people who were supposed to help Heather decorate for Prom?
4. What conclusions can you draw based on the response to Melinda's "thread" on the bathroom stall door?
5. Where does Melinda go on her first bike ride?
6. What happened between Rachel and Andy at the prom?
7. The first time Melinda was raped she couldn't find her voice and she couldn't find the inner strength to fight back. How does she react when Andy attempts to rape her again?

ANSWER KEY SHORT ANSWER STUDY GUIDE QUESTIONS – *Speak*

Assignment #1
<u>Welcome to Merryweather High-Dinner Theater</u>

1. What happens in the cafeteria that makes Melinda run out of the lunchroom?
    A lump of mashed potatoes and gravy hits Melinda in the center of her chest. Some basketball players are goofing off, and she got caught in the crossfire. Everyone in the lunchroom sees it and laughs, causing Melinda to bolt from the lunchroom to avoid more embarrassment.

2. How is Melinda's attitude in her art class different from her attitude towards her other classes?
    The title that comes before her description of art class is "sanctuary." This implies that Melinda feels like her art class is a place she can feel safe. This is the one class during the day that interests her. She actually attempts to do her work and pays attention in class.

3. Describe the globe Mr. Freeman uses in his class, and explain how he uses it.
    The globe that Mr. Freeman uses is missing half of the northern hemisphere. He used to let his daughters kick it around when it was raining and they couldn't go outside to play. After kicking it, there became a large hole in it. Mr. Freeman tells the class is could be inspiring for them, with a million creative ways to work with the globe. He then uses it to give an assignment. He has each student take a slip of paper out of the globe. Each slip has a word that the student has to draw or depict by the end of the term. This depiction of the word has to evoke emotion, which makes the assignment very challenging.

4. Describe how the students at Merryweather High School treat Melinda. Give at least three examples from the text to support your answer.
    The students at Melinda's high school all ignore her and glare at her. No one will sit with her in class or at lunch. People bump into her in the hallways and knock her books over. Others openly curse her and tell her how mad they are at her. She is also pushed down three rows of bleachers at the pep rally.

5. How does Melinda's family communicate?
    Melinda's parents work late all the time, so her family communicates mainly through sticky notes left on the kitchen counter.

6. Describe Melinda's new hiding place at school.
    Melinda finds an abandoned janitor's closet in the senior wing of the school. The closet smells bad, has dead bugs, and is a little cluttered, but there is an old armchair and a desk. Melinda plans on cleaning up the closet and sneaking in a blanket and some potpourri. She wants to use this closet as a place to hide from other people and skip class.

7. What did Melinda do that has caused everyone in school to hate her?
   Melinda called the cops at Kyle Rodgers's party at the end of the summer.

Assignment #2
Blue Roses- Third Marking Period
1. What book does Melinda read on Halloween instead of going trick-or treating?
   Melinda's parents tell her she is too old to go trick-or treating so she reads *Dracula* in her room instead.

2. Describe Melinda's lips.
   Melinda's lips are covered in scabs. She is constantly chewing on them, making them bleed and look gross.

3. The school has decided to change the school mascot four times. List all three of the previous mascots that have been used and explain why the school felt they had to change each one.
   - The Merryweather Trojans- changed because it didn't send a strong abstinence message
   - The Merryweather Devils- changed after Halloween; didn't want the school image to be so negative
   - The Merryweather Tigers- the ecology club fought because it showed a "shocking disrespect: for an endangered species

4. What author has the school district banned from the school library?
   The school district has removed all books by Maya Angelou from the library. They also removed her poster, which is how Melinda got it.

5. Explain why David Petrakis walks out of Mr. Neck's class. Give enough details in your answer to fully explain why he is upset.
   Mr. Neck is upset that his son, who is white, got passed up for a job as a firefighter. Mr. Neck thinks it is reverse discrimination and says that America should have closed her borders in 1900. He begins a class-wide debate about immigration with students arguing both sides. One student expresses an opinion opposite of Mr. Neck by pointing out that maybe his son did not get the job because someone was better than him, not because of race at all. Mr. Neck is outraged and tells the kid he is not allowed to talk and then ends the debate. David Petrakis stands up to Mr. Neck and says that if it is a debate then all opinions should be heard, not just the ones that agree with what Mr. Neck thinks. Mr. Neck tells him he is the teacher and can decide who talks in a debate, then tells him to sit down. David holds his ground, accusing the teacher of being racist and quoting his own constitutional rights. Mr. Neck threatens him to sit down or go to the principal, David looks at the flag and then leaves.

6. Describe the art project Melinda makes with the turkey bones.
   The final version of the art project Melinda makes from left over turkey bones is a museum-like display of the turkey bones, placed together like a skeleton of the dead bird. There is a Barbie doll head on top of the turkey, with tape over her mouth and a fork and knife used as legs.

7. Why does David Petrakis video his social studies class?
   David's parents have hired a lawyer over what happened in class a few days ago. They are saying that Mr. Neck is incompetent and violates civil rights. David now video tapes class to catch future incidents on tape.

8. Why is Melinda so overwhelmed by the Christmas gift her parents gave her?
   Melinda's parents say they noticed she has begun drawing and so they give her a sketch pad and some charcoal pencils. Melinda is overwhelmed by this gift because her parents actually noticed that she was doing something. She is moved that her parents have noticed something in her life because she feels like they ignore her and never notice anything. She begins to cry from being so overwhelmed with emotion and wants to tell them everything that has been going on, but they get up and leave the room.

9. Who is the "IT" Melinda refers to?
   "IT" is Andy Evans.

Assignment #3
Death of the Wombat- My Report Card

1. What animal does Melinda compare herself to whenever she sees Andy?
   Melinda's compares herself to a bunny that freezes when it gets scared and then bolts.

2. What reason does Heather give Melinda for no longer sitting with her at lunch?
   Heather tells Melinda that they are too different. She doesn't want to be friends with someone who is depressed and weird, so they can no longer sit together at lunch.

3. Who gives Melinda a Valentine card on Valentine's Day?
   Heather sticks a Valentine's card in Melinda's locker. The card thanks her for understanding about them not being friends anymore and also contains the friendship necklace that Melinda gave her for Christmas.

4. Who is the one person that Melinda speaks to?
   Mr. Freeman, the art teacher, is the one person that Melinda talks to when he gives her a ride home.

5. Melinda takes inspiration from Picasso in her tree drawing. What unique artistic method does she take from him and apply to her tree?
   Melinda discovers that Picasso used cubism. She is inspired and applies cubism to her tree drawing.

6. What is Melinda's reward for going to all her classes for an entire week?
    Melinda gets to pick out new clothes from the department store her mother works at as a reward for not skipping class. Melinda is happy to get some clothes that fit, but upset they have to come from her mother's store.

7. Describe Melinda's conflict over the party David Petrakis is throwing at his house to celebrate the basketball victory.
    Part of Melinda really wants to go. She sort of likes David and she has no other friends so she is happy to be included in his invitation to a party. However, the other part of her is terrified. She tells herself that he could be lying about his parents being there and it could get out of hand, plus she has a bad track record with parties so far. She ends up telling him no, but is torn the whole way home - part of her wishing she had gone for some fun and the other part hiding in her fear.

8. What was Melinda's initial reaction to Andy when he approached her at the party?
    At first, Melinda is very excited that a very cute boy is flirting with her. She wishes that Rachel were there to see this cute guy flirting with her. She can't believe how great he looks with his tan and perfectly toned muscles. She begins to think that she will start the school year off with a boyfriend, someone to watch over her and protect her.

Assignment #4
Exterminators- Oprah, Sally Jessy, Jerry, and Me
1. Why does the PTA want to change the hornet mascot to something else?
    The PTA has heard a cheer using the word "horny" and feels like the stinger on the cheerleaders' uniforms is provocative.

2. How does Ivy offer Melinda help?
    Ivy shows Melinda how to make her trees look more realistic. She shows her how to vary the sizes of the leaves and how to layer them.

3. Why does Melinda's dad always take a taxi to the airport?
    When Melinda was little she used to dream about being a princess who had been adopted. She thought that one day her real parents, the king and queen, would send a limo to pick her up. At age seven, a limo comes to pick her dad up and take him to the airport, and she begins to cry and panic, thinking the limo is coming to take her away. Since then, her dad has always used a taxi to go to the airport.

4. Using details from the text, describe Melinda's conflict over whether or not to warn Rachel/Rachelle about Andy.
    Melinda is torn over whether she should warn Rachel/Rachelle about Andy. On one hand, Rachel/Rachelle has been horrible to her all year and deserves whatever happens to her. On the other hand, the two have been friends since a young age, despite what has happened this year. Melinda feels like no one should have to go through what she did, however she knows that getting Rachelle to believe her won't be easy either.

5. How does Melinda warn Rachel about Andy?
    Melinda writes her an anonymous note using her left hand to try to disguise her writing. She tells her to be careful with Andy and tells her that she has heard he tried to attack a ninth grader. She then mails the note to Rachel's house.

6. What does Melinda do to get put in in-school suspension?
    Melinda is required to present her report on the suffragettes to the class. Since she does not want to speak, she makes copies of the report and distributes them to the class. She then writes on the board that she has the right to not speak. This upsets Mr. Neck, who sends her to the office, landing her in MISS once again.

7. What event prompts Melinda to go home, crawl in her closet, bury her face in her clothes, and scream at the top of her lungs?
    Melinda was working in the art room after school. She was there alone until Andy walked in. He sat on the table just inches from her and began to talk to her. She was frozen and couldn't say or do anything. He leaves a few minutes later once Ivy and Rachelle walk in, but it is just too much for Melinda. He was too close and she was scared he would hurt her again. She ran home to her own closet and buried her face in her clothes and began to scream.

Assignment #5
Real Spring-Final Cut
1. What happens at school that ruins Melinda's shirt?
    Melinda is sitting next to Ivy in art class. Ivy has four uncapped markers in her hair (stuck in a bun). When Melinda stands up, Ivy turns her head and gets marker streaks all over her shirt. It was an accident, and the stains never come out.

2. Describe how students decorate the bathroom stall doors.
    The students write all over the bathroom stall doors. Melinda observes that it is almost like a conversation board for students, with mean comments, phone numbers, and continuing conversation threads.

3. What has happened to the people who were supposed to help Heather decorate for Prom?
    Heather is a part of "the Marthas," and she is supposed to help the girls in the group who are juniors decorate for the Prom. However, the other girls get mono and can't help her decorate. Heather is then desperate to get Melinda to help.

4. What conclusions can you draw based on the response to Melinda's "thread" on the bathroom stall door?
    After a short amount of time there are already tons of responses to Melinda's note regarding "Guys to Stay Away From" on the bathroom stall door. Students from the school have responded with other warnings, some even saying that someone should tell the cops about him. This shows that other girls in the school have been victims of Andy Evans as well. One can deduce that others have been raped by him or have had to push him away after quick sexual advances.

5. Where does Melinda go on her first bike ride?
    Melinda goes to the place where she was raped.

6. What happened between Rachel and Andy at the prom?
    Andy was all over Rachel at the prom while they were dancing. She pushed him off but he wouldn't stop trying to touch her all over. She then confronted him about raping Melinda and stomped off to join her other friends. She almost slapped him, but instead left him humiliated from being dumped by a freshman.

7. The first time Melinda was raped she couldn't find her voice and she couldn't find the inner strength to fight back. How does she react when Andy attempts to rape her again?
    When Andy corners Melinda in the closet and attempts to rape her again, she is silent at first. She is terrified and feels like she can't do anything. However, she sees her poster of Maya Angelou and knows that she must do something. She throws small things at him, but he isn't bothered by her feeble attempts to get free. Even though she is pinned down by Andy, she is able to break the mirror with her turkey art project and pick up a shard of glass. Once she has the piece she makes one drop of blood come from Andy's neck and then tells him "No" once again. At that point she is saved by the lacrosse team who heard the struggle and came to see what happened. Melinda was able to fight off Andy on her own and to find her strength.

STUDY GUIDE/QUIZ QUESTIONS - *Speak*
Multiple Choice Format

Assignment #1
Welcome to Merryweather High-Dinner Theater
1. What happens in the cafeteria that makes Melinda run out of the lunchroom?
    A. Melinda trips on her way to the table, and everyone laughs.
    B. No one will let her sit with them, so she has no where to go.
    C. Some guys are goofing off, and she accidentally gets hit with mashed potatoes.
    D. The school lunch makes her feel sick.

2. How is Melinda's attitude in her art class different from her attitude towards her other classes?
    A. She thinks it is a joke and the assignments are stupid. She thinks Mr. Freeman is crazy and wishes she were in social studies instead.
    B. She feels safe in the art room. This is the only class where she pays attention and does her work.
    C. She thinks it is boring. She skips this class as much as possible.
    D. She thinks it is scary since Mr. Freeman is mean and yells at the class.

3. Describe the globe Mr. Freeman uses in his class and explain how he uses it?
    A. The globe has a huge hole in it and serves as a unique inspiration to the class. Mr. Freeman uses it to give out the semester assignment to all his students.
    B. The globe has a huge hole in it and needs to be thrown away. While at the dumpster, students find a hurt animal and try to help it.
    C. The globe has a huge hole in it from Mr. Freeman's getting angry. He was arguing with the social studies teacher and kicked it on his way out.
    D. The globe was a gift from a student Mr. Freeman had three years ago. This student went on to win several art contests and made Mr. Freeman very proud.

4. Describe how the students at Merryweather High School treat Melinda.
    A. She is a fairly popular girl. There are several guys interested in dating her, and she has lots of friends.
    B. She is hated by everyone. They push her down the bleachers, knock her books over, and say rude things to her in the halls.
    C. She is viewed as a really caring person. She volunteers to throw a teacher appreciation party, tutors other students after school, and is involved in several clubs.
    D. She is viewed as an athletic girl. She is the star of the tennis team and runs cross country.

Speak Multiple Choice Questions for assignment 1 page 2

5. Melinda's parents work all the time and are rarely home. How does her family communicate?
    A. They send each other text messages when they have something important to say.
    B. They email each other every day with an update on what's going on.
    C. They call each other's cell phone to leave a voicemail.
    D. They leave sticky notes on the kitchen counter.

6. Where is Melinda's new hiding place at school?
    A. In an abandoned janitor's closet
    B. Under the bleachers in the gym
    C. In the first stall of the girl's bathroom
    D. In the back of the library where few ever go

7. What did Melinda do that has caused everyone in school to hate her?
    A. She complained to the principal and got the school mascot changed.
    B. She let the school down in the state tennis tournament, causing the school to lose to their rival.
    C. She called the cops at a huge party over the summer.
    D. She told the principal that the most popular boy at school had drugs, which led to his arrest.

Speak Multiple Choice Questions for assignment 2

Assignment #2
Blue Roses-Third Marking Period
1. What does Melinda do on Halloween instead of going trick-or treating?
    A. She reads *Dracula*.
    B. She works on her Spanish homework.
    C. She watches scary movies and eats junk food all night.
    D. She sneaks out of the house to meet up with Heather.

2. Describe Melinda's lips.
    A. Her lips are perfectly shaped. All the other girls wish they had lips shaped like hers.
    B. Her lips are always swollen. She is constantly chewing on them, making them swell up.
    C. Her lips always have a new color of lipstick on them. She enjoys mixing the colors to create a new look each day.
    D. Her lips are covered in scabs. She is constantly chewing on them, making them bleed and look gross.

3. The school has decided to change the school mascot four times. Which of the following is NOT an explanation for why it has been changed?
    A. The Merryweather Devils—changed after Halloween because of the negative image
    B. The Merryweather Trojans— changed because it didn't send a strong abstinence message
    C. The Merryweather Bees—changed because too many people were allergic to them
    D. The Merryweather Tigers—changed because of the disrespect towards an endangered species

4. What author has the school district banned from the school library?
    A. William Shakespeare, author of several plays, including *Romeo and Juliet*
    B. Maya Angelou, poet and author of *I Know Why the Caged Bird Sings*
    C. J. K. Rowling, author of the Harry Potter series
    D. S. E. Hinton, author of several novels, including *The Outsiders*

5. Why does David Petrakis walks out of Mr. Neck's class?
    A. David is used to getting straight A's. Mr. Neck gives him a "D" on his social studies project, when David asks him why, Mr. Neck tells him it just wasn't that good and offers no other explanation, making David so angry he storms out of the room.
    B. Mr. Neck makes a racist comment about Native Americans in his lecture. David gets upset because he is part Native American. When Mr. Neck does it again, he tells him he will not sit in a room with a racist teacher—and walks out.
    C. David is a small kid who is sometimes picked on by the more popular students. One of the football players sits behind him in class and keeps poking him and kicking his seat. David turned to ask him to stop and got in trouble for talking. He tries to explain to Mr. Neck what happened but Mr. Neck he won't listen. When he gets poked again he gets up and walks out of class.

Speak Multiple Choice Questions for assignment 2 page 2

    D. Mr. Neck feels that there is reverse discrimination against white people with so many immigrants in the country. David makes a point that Mr. Neck disagrees with and gets cut off. Mr. Neck ends the discussion because he didn't like it that someone disagreed with him. David stands up for his right to voice his opinion and then leaves.

6. Describe the art project Melinda makes with the turkey bones.
    A. The turkey bones are arranged in a huge pile with a Barbie doll head in the middle. Forks and knives are arranged around the pile like they are attacking the bones.
    B. The turkey bones are arranged to look like a house. There is a Barbie doll standing in the kitchen with tape over her mouth. Forks and knives look like they are attacking the doll.
    C. The turkey bones are arranged to look like a tree. There is a Barbie doll lying on a chair made from forks and knives under the tree.
    D. The turkey bones are arranged to look like a skeleton. There is a Barbie doll head on top of the turkey with a piece of tape over her mouth. A fork and knife are used as legs.

7. Why does David Petrakis video his social studies class?
    A. He is part of an international program where students film their classes and send copies to kids in other parts of the world.
    B. He is making a video yearbook to sell at the end of the year.
    C. He wants to capture future incidents of Mr. Neck behaving inappropriately in class.
    D. He wants to help his best friend who got in a car accident and can't come to school. David videos the lessons and takes assignments to his friend's house so he doesn't fall behind.

8. Why is Melinda so overwhelmed by the Christmas gift her parents gave her?
    A. There is no art store in her town. She knows that her parents had to drive at least an hour away to get her this gift.
    B. The art supplies her parents bought are really expensive, and she is overwhelmed that they spent so much money on her gift even though she has been doing poorly in school.
    C. Melinda feels like her parents never notice her. This gift shows they have been paying attention to her new interest in art and wanted to give her something she would love.
    D. Melinda has been begging for art supplies for two months now. She is shocked and excited that her parents finally listened.

9. Who is the "IT" Melinda refers to?
    A. Heather, her ex-best friend
    B. Andy Evans, a boy she hates
    C. The principal who keeps calling her parents about her low grades
    D. The guidance counselor who keeps trying to get her to talk

Speak Multiple Choice Questions for assignment 2 page 3

10. "My parents are arguing. Not a rip-roarer. A simmering argument, a few bubbles splashing on the stove," is an example of what literary element?
    A. Tone
    B. Climax
    C. Exposition
    D. Metaphor

Multiple Choice Questions for assignment 3

Assignment #3
Death of the Wombat- My Report Card
1. What animal does Melinda compare herself to whenever she sees Andy?
	A. She compares herself to a rabbit that freezes when it gets scared and then runs away.
	B. She compares herself to a lion that roars when it feels threatened.
	C. She compares herself to a cat that prepares its claws to attack.
	D. She compares herself to a turtle that hides inside its shell when scared.

2. What reasons does Heather give Melinda for no longer sitting with her at lunch?
	A. Heather tells Melinda she is moving to New Jersey with her mom since her parents are getting a divorce. The will no longer be able to sit together since she will be a new school.
	B. Heather tells Melinda she is dating her ex-boyfriend and it would be too awkward for them to sit together at lunch.
	C. Heather tells Melinda she doesn't want to be friends with someone who is depressed and weird, so they con no longer sit together at lunch.
	D. Heather tells Melinda she has been diagnosed with cancer and will have to start treatment next week. She will be gone from school for quite a while, leaving Melinda alone at lunch.

3. Who gives Melinda a Valentine card on Valentine's Day?
	A. David Petrakis
	B. Heather
	C. Her parents
	D. A secret admirer

4. Melinda only really speaks to one person. Who can get her to open up and have a real conversation?
	A. Mr. Freeman, her art teacher
	B. Ms. Keen, her biology teacher
	C. The guidance counselor
	D. Her mom

5. Melinda is struggling with her art project. She can't figure out how to make a tree show emotion, so she looks to Picasso for inspiration. What unique artistic method does she take from him and apply to her tree?
	A. Abstract Expressionism
	B. Impressionism
	C. Realism
	D. Cubism

Speak Multiple Choice Questions for assignment 3 page 2

6. What is Melinda's reward for going to all her classes for an entire week?
    A. She gets to go to a party at David Petrakis's house.
    B. She gets to buy new clothes at the department store where her mother works.
    C. She gets to buy new art supplies.
    D. She gets to decide where her family goes for dinner to celebrate.

7. Why is Melinda's hesitant to go to the party David Petrakis is throwing at his house to celebrate the basketball victory?
    A. She is scared she won't know very many people and that she won't have anyone to talk to.
    B. She is nervous that if she goes David will know she likes him. She has never been kissed before and worries David might try to kiss her if she goes.
    C. She is scared his parents might not be there and the party will get out of hand.
    D. She has a big project due in her social studies class. If she doesn't do a good job on it, she will fail the semester and have to go to summer school.

8. What is Melinda's initial reaction to Andy when he approached her at the party last summer?
    A. She thinks he was ugly and felt like she could find someone better to date. Plus, she already had a crush on someone else.
    B. She is really excited because someone so cute was flirting with her. She thinks she will begin high school with a boyfriend to watch over her.
    C. She is worried because she knows her friend Rachel likes him. She doesn't want Rachel to be mad that Andy was flirting with her instead.
    D. She is relieved to finally see someone she knows at the party. Even though they aren't good friends, she is happy to have someone to talk to.

Speak Multiple Choice Questions for assignment 4

Assignment #4
Exterminators-Oprah, Sally Jessy, Jerry, and Me
1. Why does the PTA want to change the hornet mascot to something else?
    A. The costume for the mascot is too expensive, and there is not enough money in the budget to allow for it.
    B. Too many people think the hornet is not an aggressive mascot and would like to have it changed to something more intimidating.
    C. The PTA thinks that the new hornet cheer is too provocative and doesn't like the way cheerleaders are wearing a stinger.
    D. Too many parents have complained that their children are allergic hornets and don't want a mascot that represents an animal that could kill their children.

2. How does Ivy offer Melinda help?
    A. Ivy helps Melinda make her trees look more realistic by varying the size of leaves and layering them.
    B. Ivy helps Melinda make new friends by taking her to the mall and eating lunch with her.
    C. Ivy helps Melinda conjugate several verbs for a Spanish assignment so she won't fail the class and have to take summer school.
    D. Ivy helps Melinda redecorate her room so that is doesn't look so childish. Ivy and Melinda work to make her bedroom look more like the new person she has become.

3. Why does Melinda's dad always take a taxi to the airport?
    A. Melinda's dad thinks it is bad luck to take his own car since he once got into a bad car accident on the way to the airport.
    B. Eight years ago, a fortune teller told Melinda's family that something bad would happen on a drive to the airport. Since then they have always taken a taxi.
    C. Melinda's parents think it is ridiculous to pay for parking at the airport when they could take a taxi and save money on gas and parking.
    D. Melinda used to pretend she was a princess and one day the king and queen would come get her in a limo. When a limo showed up to take her dad to the airport she was frightened thinking it was the king and queen coming to take her away, so now he takes a taxi.

4. Melinda is unsure of whether or not she should warn Rachel/Rachelle about Andy. Which of the following is NOT a reason she is nervous to tell her?
    A. Melinda has seen Andy act like a perfect gentleman with his last few girlfriends. Melinda thinks that maybe Andy learned his lesson and has changed, giving her no reason to warn Rachel/Rachelle.
    B. Rachel/Rachelle has been really mean and hurtful to Melinda all year long. Melinda is unsure of whether or not she should be nice to her or let her learn her own lesson.
    C. Melinda is worried that Rachel/Rachelle won't believe her since they haven't been friends for so long. She also thinks Rachel/Rachelle will accuse her of being jealous.
    D. Melinda is too nervous to tell someone how she knows Andy is bad. She feels embarrassed that other people might find out what happened to her.

Speak Multiple Choice Questions for assignment 4 page 2

5. How does Melinda warn Rachel about Andy?
    A. Melinda waits until after school when Rachel is getting out of the French Club meeting and tells her face to face.
    B. She sends her a text message from someone else's phone she will not know that Melinda is the one who wrote it.
    C. Melinda writes an anonymous note with her left hand to disguise her handwriting and mails in to Rachel's house.
    D. When Rachel asks to use the bathroom during Biology, Melinda sticks a note in her book, hoping she will find it later.

6. What does Melinda do to get sent to in-school suspension?
    A. Someone sees Melinda go into the janitor's closet that she has fixed up as her hideout. This person tells the principal, and Melinda gets suspended for skipping class.
    B. Melinda gets angry at Heather for making fun of her and pushes her in the hallway. Ms. Keen, the Biology teacher, sees her do this, and Melinda gets suspended.
    C. Melinda wrote an inappropriate comment on her English paper. When her teacher read it she sent Melinda to the office where she was suspended.
    D. Melinda refuses to speak and makes photocopies of her social studies presentation instead. This makes Mr. Neck angry, and she gets suspended.

7. What event prompts Melinda to go home, crawl in her closet, bury her face in her clothes, and scream at the top of her lungs?
    A. She is angry because she finds out she will have to go to summer school.
    B. Andy Evans comes into the art room when she is alone and sits very close to Melinda.
    C. Ms. Keen yells at Melinda in front of the whole class, embarrassing her in front of everyone.
    D. Melinda's guidance counselor tells her that her parents are sending her away to an alternative school setting so she can get enough credits to graduate.

Speak Multiple Choice Questions for assignment 5

Assignment #5
Real Spring-Final Cut
1. What happens at school that ruins Melinda's shirt?
    A. Rachel and Ivy are still mad at Melinda. In an argument that takes place in the bathroom, the two girls push Melinda and tear her shirt.
    B. Ivy has uncapped markers stuck in the bun of her hair. When she turns her head the markers write all over Melinda's shirt.
    C. Two basketball players are walking down the hall. One throws a blue Gatorade to his friend, but the bottle hits a locker that opens and bursts all over Melinda's shirt.
    D. Mr. Freeman is working on a painting when Melinda asks him a question. Mr. Freeman thought he was alone, and jumps at the sound of her voice, bumping into a table with open paint cans. The paint squirts all over Mr. Freeman and Melinda.

2. What is on the back of the bathroom stall doors?
    A. The school posts important news on the stall doors to make sure students know about upcoming events.
    B. The school allows advertisements to be posted on the stall doors from businesses in the community. The school sells advertisements to these businesses to raise extra money.
    C. Students write miniature conversations on the stall doors. There are phone numbers, comments about people in school, and random thoughts from students.
    D. Students paste pictures to the stall doors. There are photos of the principal and teachers that students don't like. Students draw on the photos to make funny additions.

3. What has happened to the people who were supposed to help Heather decorate for Prom?
    A. All the other girls who are supposed to help get mono and are home sick.
    B. The sophomores and juniors decide that since Heather is only a freshman they will make her do all the work.
    C. All the other girls got caught skipping school and going to the mall instead. They are all suspended and are not allowed to help.
    D. All the girls are fighting over how to decorate for Prom. No one can agree so they all go home mad, leaving Heather to do it all.

Speak Multiple Choice Questions for assignment 5 page 2

4. How does the reader know that other people in the school have been sexually harassed or raped by Andy Evans?
    A. Rachel calls the cops after Andy almost rapes her. When the police come to the school, several other girls come forward and confess that Andy did the same thing to them.
    B. Students in Melinda's English class have to write an essay on the worst event they have ever experienced. Several of the girls in the class write their essay about being sexually harassed or raped by Andy Evans.
    C. Ms. Keen overhears Andy talking with his friends in the hall. He is bragging about all the girls he has slept with and comments that the girls don't have a choice. The teacher reports it, and everyone in the school finds out.
    D. Melinda writes a note saying to stay away from Andy on the bathroom wall. After only a few days, tons of other students have responded to her comment on the stall. Several students write about their experience with Andy, alluding to other cases of sexual harassment and rape.

5. Where does Melinda go on her first bike ride?
    A. Melinda goes to the park to find inspiration for her tree project.
    B. Melinda goes to Heather's house to try to become friends again.
    C. Melinda goes to Andy's house to confront him about the rape.
    D. Melinda goes to the place where she was raped.

6. What happened between Rachel and Andy at the prom?
    A. Andy decides that it is too embarrassing for him to be dating a freshman. He breaks up with Rachel, and she causes a big scene in front of everyone.
    B. Rachel overhears Andy talking to his friends about later that night. She finds out he got a hotel room and is expecting more than she is ready to give. She sneaks off and leaves Andy at the Prom with no explanation about where she went.
    C. Andy was all over Rachel at the Prom. She kept pushing him off, but he wouldn't stop. She ends up dumping him and leaving him alone at the Prom.
    D. Andy's ex-girlfriend shows up at the Prom alone. He dances with her while Rachel is in the bathroom. Rachel comes back to see them dancing and gets angry. She yells at him and leaves Prom crying.

Speak Multiple Choice Questions for assignment 5 page 3

7. The first time Melinda was raped she couldn't find her voice and she couldn't find the inner strength to fight back. How does she react when Andy attempts to rape her again?
   A. She is unable to move or fight back. She is silent and terrified that it is happening again, but luckily the lacrosse team is walking by the closet. They are curious what all the noise is, and when they see what is happening they stop it and run to get help.
   B. She is frozen at first but then starts to scream "No." Andy covers her mouth and tries to overpower her, but Melinda uses all her strength to break a mirror and hold a shard of glass to his throat.
   C. She immediately begins to fight him off. She has thought about this moment of revenge for so long that she knows exactly what to do. Andy ends up regretting that he ever messed with Melinda.
   D. She is unable to move or fight back. All she can do is cry while Andy rapes her again. Afterwards, she is angry at herself for letting it happen again and vows to tell the cops and get Andy arrested.

ANSWER KEY - MULTIPLE CHOICE STUDY/QUIZ QUESTIONS
Speak

	Assignment 1	Assignment 2	Assignment 3	Assignment 4	Assignment 5
1	C	A	A	C	B
2	B	D	C	A	C
3	A	C	B	D	A
4	B	B	A	A	D
5	D	D	D	C	D
6	A	D	B	D	C
7	C	C	C	B	B
8		C	B		
9		B			
10		D			

PREREADING VOCABULARY WORKSHEETS

VOCABULARY Welcome to Merryweather High-Dinner Theater *Speak*

Part I: Using Prior Knowledge and Contextual Clues

Below are the sentences in which the vocabulary words appear in the text. Read the sentence. Use any clues you can find in the sentence combined with your prior knowledge, and write what you think the underlined words mean on the lines provided.

1. I don't have time to answer. The lights dim and the <u>indoctrination</u> begins.

___

2. My first class is biology. I can't find it and get my first <u>demerit</u>.

___

3. It gives me time to scan the cafeteria for a friendly face or an <u>inconspicuous</u> corner.

___

4. <u>Sanctuary</u>. Art follows lunch, like dream follows nightmare.

___

5. Principal Principal spots another <u>errant</u> student in the hall.

___

6. She's smudging mascara under her eyes to look exhausted and <u>wan</u>.

___

7. Just a <u>pseudo</u>-friend, disposable friend. Friend as accessory.

___

8. <u>Burrow</u> (chapter title)

___

9. And they cheer on our boys, <u>inciting</u> them to violence and, we hope, victory.

___

10. We are all <u>floundering</u>.

___

Speak Vocabulary Worksheet Merryweather High-Dinner Theater Continued

Part II: Determining the Meaning
    Match the vocabulary words to their dictionary definitions.

___ 1. indoctrination	A.	a sacred place offering refuge or safety
___ 2. demerit	B.	pretend; fake; false
___ 3. inconspicuous	C.	a mark against someone for misconduct
___ 4. sanctuary	D.	straying from the right course
___ 5. errant	E.	to act clumsily or in confusion
___ 6. wan	F.	teaching someone to accept an idea or principle without criticism
___ 7. pseudo	G.	to stir, encourage, or egg on
___ 8. burrow	H.	a hole or hideout animals use to take shelter; a hideout
___ 9. inciting	I.	not noticeable; invisible
___ 10. floundering	J.	dark; gloomy; pale in color; sickly; unhappy

VOCABULARY Blue Roses-Third Grading Period *Speak*

Part I: Using Prior Knowledge and Contextual Clues

Below are the sentences in which the vocabulary words appear in the text. Read the sentence. Use any clues you can find in the sentence combined with your prior knowledge, and write what you think the underlined words mean on the lines provided.

1. The Ecology Club is planning a rally to protest the "degrading of an endangered species."

_____

2. I hang out in my refurbished closet. It is shaping up nicely.

_____

3. I am protesting the tone of this lesson as racist, intolerant, and xenophobic.

_____

4. They retreat to their corners when the phone rings again.

_____

5. Caught up in the meaning, in the subjectivity of the effect of commercialism on this holiday.

_____

6. The knife and fork are obviously middle-class sensibilities.

_____

7. The revolutionary spirit is even breaking out in social studies class.

_____

8. I'm just like them—an ordinary drone dressed in secrets and lies.

_____

9. Ms. Conners did not win a lacrosse scholarship by being demure or hesitant.

_____

10. Mount Dad, long dormant, now considered armed and dangerous.

_____

Speak Vocabulary Worksheet Blue Roses-Third Grading Period Continued

Part II: Determining the Meaning
Match the vocabulary words to their dictionary definitions.

___ 1. degrading   A. to withdraw or go back
___ 2. refurbished   B. emotions or feelings
___ 3. xenophobic   C. shy, modest, coy
___ 4. retreat   D. humiliating; disgrace; dishonor
___ 5. subjectivity   E. support for radical change or innovation
___ 6. sensibilities   F. unreasonable fear or hatred of foreigners
___ 7. revolutionary   G. inactive; lying asleep; not erupting
___ 8. drone   H. decisions based on personal feelings rather than facts
___ 9. demure   I. a remote control mechanism
___ 10. dormant   J. to make clean, bright, or fresh again; renovate

VOCABULARY Death of the Wombat-My Report Card *Speak*

Part I: Using Prior Knowledge and Contextual Clues

Below are the sentences in which the vocabulary words appear in the text. Read the sentence. Use any clues you can find in the sentence combined with your prior knowledge, and write what you think the underlined words mean on the lines provided.

1. How can I shop with Mom if I don't want to talk to her? She might love it—no arguing that way. But then I'd have to wear the clothes she picked out. <u>Conundrum</u>—a three-point vocab word.

   _____

2. It's the blood of <u>imbeciles</u>.

   _____

3. The squeal of announcements no one hears—it is a <u>vespiary</u>, the Hornet haven.

   _____

4. I think we need to explore the family <u>dynamics</u> at play here.

   _____

5. Do they choose to be so <u>dense</u>?

   _____

6. It is supposed to bore us into <u>submission</u> or prepare us for the insane asylum.

   _____

7. I think this is part of his punishment for that <u>bigoted</u> crap he pulled in class.

   _____

8. Choked by weeds, rooted up by dogs, mashed by a soccer ball, or <u>asphyxiated</u> by car exhaust.

   _____

9. She looks <u>wistful</u>.

   _____

10. David doesn't bother to analyze my <u>reluctance</u>.

    _____

Speak Vocabulary Worksheet Death of the Wombat-My Report Card Continued

Part II: Determining the Meaning
    Match the vocabulary words to their dictionary definitions.

___ 1. Conundrum	A. a group of stupid or silly people
___ 2. Imbeciles	B. surrendering power to another; acting in a way of meekness
___ 3. Vespiary	C. intolerant of any other beliefs or opinions
___ 4. Dynamics	D. pensive; thoughtful in a sad way; longing yearning
___ 5. Dense	E. a nest of social wasps
___ 6. Submission	F. acting dull or slow-witted; thinking in a stupid manner
___ 7. Bigoted	G. a difficult problem; a dilemma
___ 8. Asphyxiated	H. choked; suffocated; smothered
___ 9. Wistful	I. unwillingness; resisting
___ 10. Reluctance	J. the social, intellectual, or physical forces that characterize a system or group

VOCABULARY Exterminators-Oprah, Sally Jessy, Jerry, and Me *Speak*

Part I: Using Prior Knowledge and Contextual Clues

Below are the sentences in which the vocabulary words appear in the text. Read the sentence. Use any clues you can find in the sentence combined with your prior knowledge, and write what you think the underlined words mean on the lines provided.

1. It pleads for consistency, stability.

_____

2. We are tenacious, stinging, clever.

_____

3. The last unit of the year in biology is genetics.

_____

4. Ms. Keen writes "Dominant/Recessive" on the board.

_____

5. The friendly momentum keeps Rachel/Rachelle and Andy walking.

_____

6. The best place to figure this out is my closet, my throne room, my foster home.

_____

7. I didn't want the Swedish supermodel on my conscience either.

_____

8. Instead of leaving me alone to "find my muse," he lands on the stool next to me.

_____

9. Maybe I don't want to incriminate myself.

_____

10. They buy shoes and go to the movies. Cutting-edge adult delinquency.

_____

Speak Vocabulary Worksheet Exterminators-Oprah, Sally Jessy, Jerry, and Me Continued

Part II: Determining the Meaning
    Match the vocabulary words to their dictionary definitions.

   ___ 1. Consistency   A. science of heredity and genes
   ___ 2. Tenacious   B. something that nourishes or cares for; encouraging
   ___ 3. Genetics   C. something or someone that is inspiring to an artist
   ___ 4. Recessive   D. persistent; stubborn; vicious; not easily pulled apart
   ___ 5. Momentum   E. force or speed of movement; motion
   ___ 6. Foster   F. wrongful; illegal; failure to fulfill a duty or obligation
   ___ 7. Conscience   G. the inner sense of what is right or wrong
   ___ 8. Muse   H. to make someone appear guilty in a crime
   ___ 9. Incriminate   I. going back; receding; a gene that does not produce
   ___ 10. Delinquency   J. keeping the same behavior, form, pattern, or principles

Note: The word "foster" literally means to nourish or care for. A foster home is somewhere someone is safe and cared for other than in their regular home. In the case of Melinda, the closet was a foster home in the sense that it was not her real home; it was a place of safety and a place where she could use her inner sources of nourishment to sprout.

VOCABULARY Real Spring-Final Cut *Speak*

Part I: Using Prior Knowledge and Contextual Clues
Below are the sentences in which the vocabulary words appear in the text. Read the sentence. Use any clues you can find in the sentence combined with your prior knowledge, and write what you think the underlined words mean on the lines provided.

1. The sun appears butter-yellow and so warm it coaxes tulips out of the crusty mud.

_____

2. You hunt down every person who looks vaguely familiar and get them to write in your yearbook.

_____

3. He has such a reputation. He's only after one thing.

_____

4. My mother can't believe it: a living, breathing friend on the front porch for her maladjusted daughter.

_____

5. She dumped me, banished me from even the shadows of Martha glory.

_____

6. She is talking to the village leper.

_____

7. He sets to work pruning the deadwood like a sculptor.

_____

8. Some part of me has planned this, a devious internal compass pointed to the past.

_____

9. One stomach pumped, three breakups of long term relationships, one lost diamond earring, four outrageous hotel-room parties, and five matching tattoos allegedly decorating the behinds of the senior class officers.

_____

Speak Vocabulary Worksheet Real Spring-Final Cut Continued

10. Sometimes I think high school is one long <u>hazing</u> activity.

_____

Part II: Determining the Meaning
   Match the vocabulary words to their dictionary definitions.

   ___ 1. Coaxes   A. poorly adjusted in one's social circumstances
   ___ 2. Vaguely   B. something that has been described but not proven
   ___ 3. Reputation   C. to persuade by pleading or flattery
   ___ 4. Maladjusted   D. a person who is rejected; an outcast
   ___ 5. Banished   E. not straightforward; shifty or crooked
   ___ 6. Leper   F. not clear or definite; hazy
   ___ 7. Pruning   G. how the public views or regards an individual
   ___ 8. Devious   H. to abuse newcomers with humiliating tricks and ridicule
   ___ 9. Allegedly   I. forced to leave; drive away; expelled
   ___ 10. Hazing   J. to cut off, clear, or remove

Notes: In the old days, a person with a disease called leprosy was called a **leper**. This disease could cause open sores, inflamed tissue bumps, mutilation, gangrene, and paralysis. It could be highly contagious. There was no cure. When a person got leprosy, they were sent out of the village or town to live alone or sometimes in a colony of lepers. Thus, someone who is outcast from a group is sometimes called a leper.

Pruning is the trimming of a tree or bush to remove dead limbs or make it more compact. Sometimes if a tree or bush is not doing well–has bad parts or has grown out of control with some good parts and some bad–pruning it (cutting it back) to a more compact shape will enable it to become rejuvenated.

VOCABULARY ANSWER KEY *Speak*

	Assignment 1	Assignment 2	Assignment 3	Assignment 4	Assignment 5
1	F	D	G	J	C
2	C	J	A	D	F
3	I	F	E	A	G
4	A	A	J	I	A
5	D	H	F	E	I
6	J	B	B	B	D
7	B	E	C	G	J
8	H	I	H	C	E
9	G	C	D	H	B
10	E	G	I	F	H

DAILY LESSONS

LESSON ONE

Objectives
1. To introduce the *Speak* unit
2. To distribute books, study questions, and other related materials
3. To preview the vocabulary and study questions for Assignment #1
4. To begin Assignment #1

Note: The level of your students will determine how involved you want to make this introductory activity. You may simply want students to draw a picture of a tree to display around your classroom, or you may want to have students plant their own seeds and care for the plants over the course of the unit, watching them grow like Melinda does over the course of the novel. You can tell students to bring in a package of seeds and a small pot/cup from home, or you can purchase all materials and have them for students. (*If students are bringing their own materials, be sure to have extras for those who forget.*)

Activity #1
Ask students to think about the growth of a tree. Direct students think about the entire process, from the initial planting of the seed through the end of the tree's life. Have students think about the dangers and problems that can occur over the course of the tree's life (uprooted, cut down, damage from extreme weather, diseased limbs, etc). Once you have talked in general about the life of a tree, ask students to list qualities of a tree (strong, survive almost anything, constantly growing, changes with the seasons).

Activity #2
Next, give each student a small pot or cup filled with soil. Distribute seeds to the class, explaining the proper care of the plants (water and sunlight). You may want to appoint a time each day (the first two minutes or last two minutes of class) for students to water their plants and check on the progress. Allow a few minutes for students to label their pots/cups and plant their seeds.
    Transition: Once you have discussed the characteristics and life of trees, tell students they are about to begin reading *Speak*. Show them the front of the novel, pointing out that there is a tree on the cover. Tell students that over the course of the novel several references to trees will be made. Explain to them that trees are a major symbol used through the novel and will be used to represent Melinda's growth and life.

Activity #3
Distribute the Raising Awareness project and rubric. Discuss the directions in detail. *Note: Some schools may allow for these posters to be displayed around the school to raise awareness about rape. If your school allows this, select your best posters to hang in popular areas around school. If not, display the posters in your room.*

Activity #4
Distribute the materials students will use in this unit. Explain in detail how students are to use these materials.

Study Guides   Students should read the study guide questions for each reading assignment prior to beginning the reading assignment to get a feeling for what events and ideas are important in the section they are about to read. After reading the section, students will (as a class or individually) answer the questions to review the important events and ideas from that section of the book. Students should keep the study guides as study materials for the unit test. **Review the study questions for Assignment 1 while you're looking at the study guides.**

Vocabulary   Prior to reading a reading assignment, students will do vocabulary work related to the section of the book they are about to read. Following the completion of the reading of the book, there will be a vocabulary review of all the words used in the vocabulary assignments. Students should keep their vocabulary work as study materials for the unit test. **Do Assignment 1 together orally to show students how to do the vocabulary worksheet.**

Reading Assignment Sheet   You need to fill in the Reading Assignment Sheet to let students know by when their reading has to be completed. You can either write the assignment sheet up on a side blackboard or bulletin board and leave it there for students to see each day, or you can "ditto" copies for each student to have. In either case, you should advise students to become very familiar with the reading assignments so they know what is expected of them.

Extra Activities Center   The Unit Resource Materials portion of this LitPlan contains suggestions for an extra library of related books and articles in your classroom as well as crossword and word search puzzles. Make an extra activities center in your room where you will keep these materials for students to use. (Bring the books and articles in from the library and keep several copies of the puzzles on hand.) Explain to students that these materials are available for students to use when they finish reading assignments or other class work early.

Nonfiction Assignment Sheet   Explain to students that they each are to read at least one non-fiction piece from the in-class library at some time during the unit. Students will fill out a nonfiction assignment sheet after completing the reading to help you (the teacher) evaluate their reading experiences and to help the students think about and evaluate their own reading experiences.

Books   Each school has its own rules and regulations regarding student use of school books. Advise students of the procedures that are normal for your school. Preview the book. Look at the covers, frontmatter, and index.

Activity #5
Tell students that they should read Assignment #1 prior to the next class period. Give them the remainder of this class (if time remains) to complete this assignment.

RAISING AWARENESS PROJECT

This is an individual assignment that you will need to complete at home. The goal of your project is to find a creative way to make other students more aware of the serious issue discussed in *Speak* (rape). Read over all of the instructions on this sheet and in your rubric for the highest possible grade. All projects will be presented to the class and displayed for others to see.

Project Requirements:
Each project should address the issue of rape. Your project should include at least three powerful passages from the book that address the serious nature of your issue. You will also need a minimum of eight facts regarding your issue. These facts need to be from at least two different sources, none of which can be your book. You will need to use parenthetical documentation with your passages and your facts. A Works Cited page needs to be turned in with your project as well. Each project should be creative, neat, and free of all spelling and grammar mistakes. Be sure to include graphics and pictures while keeping the assignment school appropriate.

Project Choices:
- Poster: needs to be done on poster board using markers, paint, or typed material so that it is legible from a short distance
- Mobile: should have multiple pieces of paper or some other material hanging at varying lengths; should not be done in pen or pencil
- Diorama: should be in a covered or painted box; should be colorful and three dimensional
- Quilt: should use various materials and colors to provide a contrast between pieces; should not be done in pen or pencil; should be sewn or put together in an authentic manner

RAISING AWARENESS PROJECT RUBRIC

Category	10 Points	7 Points	5 Points	0 Points	Points Assigned
Passages from the book	At least 3 powerful passages from the book that apply to the issue covered	At least 2 powerful passages from the book that apply to the issue covered	At least 1 powerful passage form the book that applies to the issue covered	No passages from the book or the passages do not pertain to the issue covered	
Facts/Research	At least 8 facts regarding the issue covered	6-7 facts regarding the issue covered	4-5 facts regarding the issue covered	3 or fewer facts regarding the issued covered	
Variety of research sources	At least 2 resources for research, aside from the book	1 resource for research, aside from the book	No additional resources for the research other than the book	No evidence of research is included in the project	
Works Cited	Accurate works cited page	A somewhat accurate works cited page	Several mistakes are present on the works cited page	No works cited page was turned in	
Parenthetical Documentation	All of the facts and passages follow correct parenthetical documentation format	Most of the facts and passages follow correct parenthetical documentation format	Some of the facts and passages follow correct parenthetical documentation format	None of the facts and passages follow correct parenthetical documentation format	
Grammar	There are no grammatical mistakes	There are few grammatical mistakes	There are several grammatical mistakes	More than half the assignment contains grammatical mistakes	
Title	Title covers the issue and is creative	Title somewhat covers the issue or is somewhat creative	Title is included but does not cover the issue or is not creative	No title is present	
Attractiveness	Exceptionally attractive in terms of neatness, design, and layout	Attractive in terms of neatness, design, and layout	Acceptably attractive in terms of neatness, design, and layout	Messy and poorly designed	
Graphics	Several graphics that relate to the issue are used	2-3 graphics are used and mostly relate to the issue	The graphics do not relate to the issue	No graphics were included	
Creativity	The student has gone above and beyond expectations, incorporating several unique ideas	The student is somewhat creative and does a good job of adding new ideas to the project	The student has done very little to be creative	No creativity is present	
				Total Points:	

LESSON TWO

Objectives
1. To review main ideas, events, and vocabulary of reading Assignment #1
2. To assess what the students know about rape and speaking out against issues, determine what the class would like to learn, and prepare a place for the students to display their knowledge as the unit progresses
3. Create a visual display that will be used as a prereading, during reading, and after reading strategy

Activity #1
Give students a few minutes to formulate answers for the study guide questions for reading Assignment #1, and then discuss the answers to the questions in detail. Write the answers on the board or overhead transparency so students can have the correct answers for study purposes.

NOTE: It is a good practice in public speaking and leadership skills for individual students to take charge of leading the discussions of the study questions. Perhaps a different student could go to the front of the class and lead the discussion each day that the study questions are discussed in this unit. Of course, you should guide the discussion when appropriate and try to fill in any gaps students may leave. The study questions could really be handled in a number of different ways, including in small groups with group reports following. Occasionally you may want to use the multiple choice questions as quizzes to check students' reading comprehension. As a short review now and then, students could pair up for the first (or last, if you have time left at the end of a class period) few minutes of class to quiz each other from the study questions. Mix up methods of reviewing the materials and checking comprehension throughout the unit so students don't get bored just answering the questions the same way each day. Variety in methods will also help address the different learning styles of your students. From now on in this unit, the directions will simply say, "Discuss the answers to the study questions in detail as previously directed." You will choose the method of preparation and discussion each day based on what best suits you and your class.

Activity #2
Review the vocabulary answers from the reading. Make sure students write down the correct answers.

Activity #3
Create a KWL chart with your class. Using construction or butcher paper, create a tree trunk with several roots extending from the tree. This can be as large or as small as you wish, depending on how much wall space you have to hang your KWL chart (this could also be done on a bulletin board space). First, ask students to brainstorm what they already know about rape, speaking up, finding a strong voice, growing as an individual, enduring hardships, etc. Write all of these ideas on the board. Once everyone has contributed, select the top six or seven responses and allow students to write them on the roots of the tree. The roots will symbolize the knowledge base "what we *know*" for the unit. Next, ask students to brainstorm "what they *want* to know" about these same topics and write the list of ideas on the board. Select several good responses and allow students to write them on the trunk of the tree. Finally, have students cut out several leaves. Place the leaves in a box or basket somewhere in the room. As the unit progresses and students learn more about the topics mention above, have them write "what they *learned*" on a leaf and attach it to the tree. This KWL chart will serve as a prereading, during reading, and after reading strategy and will get students excited to create something to decorate the room.

KWL CHART - *Speak*

What I Know	What I Want to Learn	What I Learned

LESSON THREE

Objectives
1. To review and study literary elements and terminology
2. To teach students about parenthetical documentation when using research
3. To teach students about creating a works cited page when using research
4. To preview the vocabulary and study questions for Assignment #2
5. To read Assignment #2

Activity #1
Review commonly used literary elements and terminology with your students. These terms should be in your state or county curriculum guide. A list of commonly taught terms will be included for your use if necessary. Students should already be familiar with these terms, therefore some form of review should be conducted. A matching review sheet of commonly taught terms is provided. Depending on how much time you have already spent on these terms previously in the course, and the level of your students, review as you see fit.

Activity #2
Teach students about using parenthetical documentation and creating a works cited page. On the board or overhead, define the following terms: summary, paraphrase, direct quotations, and plagiarism. Talk to your students about these terms and how they relate to research and citation of information. Briefly discuss the importance of crediting a source when conducting research or using quotations. Most English textbooks will come with worksheets and explanations of parenthetical documentation and works cited pages. Also, depending on the age and level of your students, some may already be familiar with this process. Assess how much your students know and gather the resources you will need to either teach them about parenthetical documentation and works cited pages or review the information they already know. The MLA Handbook contains several resources and reviews for teaching these skills.

Activity #3
Look ahead at the study questions and vocabulary for Assignment #2. Tell students that they should read Assignment #2 prior to the next class period. Give them the remainder of this class (if time remains) to complete this assignment.

LITERARY TERMS

1. **Plot**: series of related events that make up a story or drama
2. **Exposition**: the beginning part of a story that introduces the characters, the setting, and the basic situation; also introduces problems or conflicts
3. **Rising Action**: events after the exposition and leading up to the climax of the story
4. **Climax**: the highest point of interest or suspense in a story; exciting or tense moments, usually toward the end of the story, when we realize what the outcome of the conflict will be
5. **Falling Action**: all events after climax and leading to the resolution of a story
6. **Resolution**: the moment when all the problems are resolved, one way or another and the story is closed
7. **Setting**: the time and place of the action of a story or play
8. **Conflict**: struggle or clash between opposing characters of opposing forces
   - **External Conflict**: a character struggles against an outside force, such as another character, nature, or the standards/expectations of society/group
   - **Internal Conflict**: a struggle opposing need, desires, or emotions within a single characters mind
9. **Point of View**: vantage point from which the writer tells a story
   - **First Person**: one of the characters is telling the story using I; we only know information that the character sees, hears, or knows
   - **Third Person Limited**: the narrator, who plays no part in the story, tell us the thoughts and feelings of only one character in the story
   - **Third Person Omniscient**: the person telling the story knows everything there is to know about the character's past, present, and future
10. **Symbolism**: a person, place, or thing, that stands for itself and for something beyond itself
11. **Mood**: atmosphere or the feeling created by the writer in the story; it is often suggested by descriptive details
12. **Suspense**: a feeling of uncertainty, anxiety, or curiosity about what is going to happen next or about the outcome of events in the story
13. **Style**: some writers use simple, down to earth, or even slang words; others use ornate, official-sounding, or even flowery language
14. **Characterization**: process or act of creating, developing, or revealing the personality of a person/character in a story
    - **Indirect Characterization**: using your own judgment to decide what a character is like based on evidence given by the writer, such as looks, actions, speech, and reactions of others
    - **Direct Characterization**: author directly states a character's traits Static Character: does not change much in the course of the story
    - **Dynamic Character**: changes as a result of the story's events
    - **Flat Character**: has no depth; only has one or two traits and these can be described in a few words
    - **Round Character**: like a real person; has many character traits, which sometimes contradict one another; has many faults and virtues
15. **Theme**: central idea, message, or insight into life revealed through the story
16. **Protagonist**: main or most important character in the story; the good guy

Literary Terms Continued

17. **Antagonist**: a major character or force who opposes or is in conflict with the main character and or protagonist; the bad guy
18. **Dialogue**: a conversation between two characters that is used to reveal character and to advance the story line or action
19. **Irony**: contrast or discrepancy between expectation and reality – between what is said and what is really meant; between what is expected to happen and what really does happen; or between what appears to be true and what is really true
    - **Verbal Irony**: words are used to suggest the opposite of what is meant
    - **Dramatic Irony**: the audience or reader knows something important that the main character in the story does not know
    - **Situational Irony**: an event occurs that directly contradicts the expectations of the characters, the reader, or the audience
20. **Foreshadowing**: use of clues to hint at events that will occur later in the plot
21. **Allusion**: reference in a story to a statement, a person, a place, or an event from literature, history, religion, myth, politics, sports, science, or a pop culture
22. **Satire**: type of writing that uses humor or criticism to ridicule a group of people, humanity at large, an attitude or failing, or a social institution in order to reveal weakness and/or in hopes of improving them
23. **Tone**: attitude the writer takes toward the audience, a subject, or a character; it is conveyed through the writer's choice of words and details; can be formal, informal, serious, playful, bitter, ironic, etc
24. **Diction**: a writer's or speaker's choice of words; using different types of words depending on the audience being addressed, the subject being discussed, and the effect or mood trying to be produced
25. **Dialect**: way of speaking that is characteristic of a particular region or group
26. **Anecdote**: brief story about an interesting, amusing, or strange event used to entertain or to make a point
27. **Flashback**: section of a literary work that interrupts the sequence of events to relate an event from an earlier time
28. **Alliteration**: repeating the same beginning sound of closely linked words
29. **Simile**: comparing 2 things to add meaning using connective words ("like" or "as")
30. **Metaphor**: comparing 2 things to add meaning without using connective words
31. **Personification**: when an animal, object, or force is given human personality traits
32. **Hyperbole**: overstatement or exaggeration for special effect
33. **Imagery**: descriptive language that appeals to any of the senses

LITERATURE TERMS REVIEW

Matching: Match the literary term with the correct definition

A.	Rising Action	J.	Verbal Irony	S.	Setting
B.	Falling Action	K.	Dramatic Irony	T.	Flat Character
C.	Symbolism	L.	Situational Irony	U.	Round Character
D.	Exposition	M.	Flashback	V.	Indirect Characterization
E.	Climax	N.	Foreshadowing	W.	Direct Characterization
F.	Resolution	O.	Mood	X.	Simile
G.	Static Character	P.	Protagonist	Y.	Metaphor
H.	Dynamic Character	Q.	Antagonist	Z.	Personification
I.	Theme	R.	Suspense	aa.	Imagery

____ 1. an event occurs that directly contradicts the expectations of the characters, the reader, or the audience

____ 2. the beginning part of a story that introduces the characters, setting, and basic situation

____ 3. events after climax and leading to the resolution of a story

____ 4. main or most important character in the story; the good guy

____ 5. author directly states a character's traits or what kind of person the character is

____ 6. a feeling of uncertainty, anxiety, or curiosity about what is going to happen next or about the outcome of events

____ 7. a person, place, or thing, that stands for itself and for something beyond itself

____ 8. all events after the exposition and leading up to the climax of the story

____ 9. use of clues to hint at events that will occur later in the plot

____ 10. descriptive language that appeals to any of the senses

____ 11. a major character or force who opposes or is in conflict with the main character and or protagonist; the bad guy

____ 12. atmosphere or the feeling created by the writer in the story

____ 13. central idea, message, or insight into life revealed through the story

____ 14. character that is like a real person with many character traits; has many faults and virtues

____ 15. section of a literary work that interrupts the sequence of events to relate an event from an earlier time

____ 16. character who changes as a result of the story's events

____ 17. when an animal, object, or force is given human personality traits

____ 18. the audience or reader knows something important that the main character in the story does not know

____ 19. the moment when all the problems are resolved, one way or another, and the story is closed

____ 20. the highest point of interest or suspense in a story

____ 21. character who does not change much in the course of the story

____ 22. comparing 2 things to add meaning without using connective words

____ 23. words are used to suggest the opposite of what is meant

____ 24. using your own judgment to decide what a character is like based on evidence given by the writer

____ 25. the time and place of the action of a story

____ 26. comparing 2 things to add meaning using connective words ("like" or "as")

____ 27. character with no depth and hardly any character traits

LITERATURE TERMS REVIEW ANSWER KEY

Matching: Match the literary term with the correct definition.

A.	Rising Action	J.	Verbal Irony	S.	Setting
B.	Falling Action	K.	Dramatic Irony	T.	Flat Character
C.	Symbolism	L.	Situational Irony	U.	Round Character
D.	Exposition	M.	Flashback	V.	Indirect Characterization
E.	Climax	N.	Foreshadowing	W.	Direct Characterization
F.	Resolution	O.	Mood	X.	Simile
G.	Static Character	P.	Protagonist	Y.	Metaphor
H.	Dynamic Character	Q.	Antagonist	Z.	Personification
I.	Theme	R.	Suspense	aa.	Imagery

L   1. an event occurs that directly contradicts the expectations of the characters, the reader, or the audience

D   2. the beginning part of a story that introduces the characters, setting, and basic situation

B   3. events after climax and leading to the resolution of a story

P   4. main or most important character in the story; the good guy

W   5. author directly states a character's traits or what kind of person the character is

R   6. a feeling of uncertainty, anxiety, or curiosity about what is going to happen next or about the outcome of events

C   7. using a person, place, or thing, that stands for itself and for something beyond itself

A   8. all events after the exposition and leading up to the climax of the story

N   9. use of clues to hint at events that will occur later in the plot

aa   10. descriptive language that appeals to any of the senses

Q   11. a major character or force who opposes or is in conflict with the main character and or protagonist; the bad guy

O   12. atmosphere or the feeling created by the writer in the story

I   13. central idea, message, or insight into life revealed through the story

U   14. character that is like a real person with many character traits; has many faults and virtues

M   15. section of a literary work that interrupts the sequence of events to relate an event from an earlier time

H   16. character who changes as a result of the story's events

Z   17. when an animal, object, or force is given human personality traits

K   18. the audience or reader knows something important that the main character in the story does not know

F   19. the moment when all the problems are resolved, one way or another, and the story is closed

E   20. the highest point of interest or suspense in a story

G   21. character who does not change much in the course of the story

Y   22. comparing 2 things to add meaning without using connective words

J   23. words are used to suggest the opposite of what is meant

V   24. using your own judgment to decide what a character is like based on evidence given by the writer

S   25. the time and place of the action of a story

X   26. comparing 2 things to add meaning using connective words ("like" or "as")

T   27. character with no depth and hardly any character traits

LESSON FOUR

<u>Objectives</u>
1. To review main ideas, events, and vocabulary of reading Assignment #2
2. To allow to students to express personal opinions and use creativity in a writing assignment modeled after an assignment given to Melinda in *Speak*
3. To enhance students' writing ability

Activity #1
Have students answer the study guide questions for reading Assignment #2 as previously directed.

Activity #2
Review the vocabulary answers from the reading. Make sure students write down the correct answers.

Activity #3
In *Speak*, Melinda is given a writing assignment from her English teacher to write "The Best Lost Homework Excuse Ever" in five hundred words. Students should have just completed this portion of the reading and will have the same assignment that Melinda and her classmates did. You can vary the amount of words required based on the ability level of your students, but the essay should be a creative story that goes into great detail about the excuse. Distribute Writing Assignment #1 and use the attached rubric to give feedback to your students.

WRITING ASSIGNMENT #1 - *Speak*
Writing to express personal opinions

PROMPT
Melinda and her classmates had to write an essay in their English class. The assignment was to write an essay about "The Best Lost Homework Excuse Ever." Your assignment is to write a creative and detailed essay about the best lost/late homework excuse you can think of.

PREWRITING
Think about all the excuses you or your friends have ever given your teacher for lost or late homework. What excuses were the most creative? What excuses worked or made the teacher laugh? Lost and late homework is often the topic of television shows and movies as well. Think about the shows you watch and some of the creative excuses you've seen on screen. What excuses were the most detailed?

Write down some possible ideas for your excuse. Remember, a good, creative excuse will have several steps to it. Think about the entire course of events that led up to your homework being late or lost.

DRAFTING
This essay should be written in first person. It will have an introductory paragraph where you will want to mention the assignment that is late and the class/teacher you are explaining the excuse to.

The body of your essay should include several paragraphs explaining all the details of the excuse. When a new problem occurs, begin a new paragraph. Try to separate the steps that lead up to the homework being lost or late into their own paragraph. Remember, a good excuse will contain several attempts at getting the assignment done, or finding the homework, but with a new problem that prevents this from happening in each paragraph.

Your essay should have a conclusion that contains the end to your story. This should include the last step—your arrival at school. You will also want to mention the assignment, teacher, and class again to remind the reader of what it was that was late.

PROMPT
When you finish the rough draft of your composition, ask a student who sits near you to read it. After reading your rough draft, he/she should tell you what he/she liked best about your work, which parts were difficult to understand, and ways in which your work could be improved. Reread your paper considering your critic's comments, and make the corrections you think are necessary. Ask your classmate what he/she thought of each of the characters/events you chose for your assignment.

PROOFREADING
Do a final proofreading of your paper double-checking your grammar, spelling, organization, and the clarity of your ideas.

WRITING EVALUATION FORM
Speak

Name _____ Date _____

Writing Assignment #_____ Grade_____

Circle One for Each Item:

Composition	Excellent	Good	Fair	Poor
Style	Excellent	Good	Fair	Poor
Grammar	Excellent	Good	Fair	Poor
Spelling	Excellent	Good	Fair	Poor
Punctuation	Excellent	Good	Fair	Poor
Legibility	Excellent	Good	Fair	Poor

Strengths:

Weaknesses:

Comments/Suggestions:

LESSON FIVE

Objectives
1. To preview the vocabulary and study questions for Assignment #3
2. To read Assignment #3
3. To discuss symbolism and push students to recognize symbolism
4. To allow students to create their own projects that contain symbolism
5. To allow students to use art as a from of expression as Melinda does in the novel

Activity #1
Melinda saves the bones from the Thanksgiving turkey and uses them to create a powerful art project. Melinda combines odds and ends she finds in a junk box Mr. Freeman has in his classroom to create a project with a lot of symbolism. By combining several odd items, she creates a piece of art that expresses her pain. Mr. Freeman and Ivy both look at the project and comment on the symbolism of each item used in the project. The turkey bones, Barbie head, fork, knife, plastic palm tree, and tape all stand for something. Instruct your students to bring in five or six random pieces of "junk" from home. Clean out any junk drawers you may have and bring those items into class. If possible, stop at a garage sale or Goodwill to purchase small, and cheap, items to bring in for this project. Items should have no method to them and be random items of all sorts.

Reread the chapter entitled "Wishbone" with your class. Talk about how each piece of Melinda's project stood for something. Discuss how she uses items to create a meaning, each item being a symbol for something. Also discuss how colors can often be a symbol and review the meaning behind several colors with your students.

Next place the items you brought in at the front of the classroom. Put students into groups of three or four and allow them a few minutes to combine their items and collect any additional items they may need. Provide them with tape, glue, and colored construction paper. Tell students they need to create an art project similar to Melinda's project that hold meaning. Remind them to select a color of paper to do the project on that will add meaning to the project. Once they have constructed the art project, ask groups to write a short paragraph outlining what each piece symbolizes.

After students have finished, go around the room and allow groups to show off their art project. Urge them to talk about what each piece symbolizes to them and what message the entire project gives off to viewers. If possible, display these projects around your room.

Activity #2
Look ahead at the study questions and vocabulary for Assignment #3. Tell students that they should read Assignment #3 prior to the next class period. Give them the remainder of this class (if time remains) to complete this assignment.

LESSON SIX

Objectives
1. To review main ideas, events, and vocabulary of reading Assignment #3
2. To preview study questions and vocabulary for Assignment #4
3. To read Assignment #4
4. To evaluate students' oral reading

Activity #1

Have students answer the study guide questions for reading Assignment #3 as previously directed.

Activity #2

Review the vocabulary answers from the reading. Make sure students write down the correct answers.

Activity #3

Look ahead at the study questions and vocabulary for Assignment #4.

Activity #3

Have students read Assignment #4 of *Speak* out loud in class. You probably know the best way to get readers with your class; pick students at random, ask for volunteers, or use whatever method works best for your group. If you have not yet completed an oral reading evaluation for your students, this would be a good opportunity to do so. A form is included with this unit for your convenience.

ORAL READING EVALUATION
Speak

Name _____ Class _____ Date _____

SKILL	EXCELLENT	GOOD	AVERAGE	FAIR	POOR
FLUENCY	5	4	3	2	1
CLARITY	5	4	3	2	1
AUDIBILITY	5	4	3	2	1
PRONUNCIATION	5	4	3	2	1
_____	5	4	3	2	1
_____	5	4	3	2	1

TOTAL_____GRADE_____

COMMENTS:

LESSON SEVEN

Objectives
1. To bring ideas from the book into real life
2. To inform students about rape and how to speak up when it comes to important issues

Activity #1
We have set this day aside for a guest speaker. Invite one or more of the following people from your community to speak to your class:

    Self-defense instructor to discuss ways to avoid situations of danger
    Rape victims
    Police officer that has dealt with rapes
    Doctor that can discuss rape and what to do
    Spokesperson from a rape crisis center
    Any other person that could educate your students on rape

Divide your class time according to how many speakers you're able to acquire. Remember to allow time for students to ask questions. Let each speaker know how much time he/she will have for the presentation. Allow for time at the end of the class for students to make connections with what they have learned from the speakers with what they have read in *Speak*.

Follow Up: Be sure you and your students write thank you notes to each of your guests. At the very least, get a thank you card for each guest and have each of your students sign it (with any personal responses, if there is room).

LESSON EIGHT

Objectives
1. To give students the opportunity to practice writing to persuade
2. To improve students' writing ability
3. To connect ideas facts learned from the speaker and book with a real-life situation

Activity
Since students have had a guest speaker discussing rape and have read enough about Melinda's rape and her difficulty talking about it afterwards; they should now be ready to write a persuasive essay. This persuasive essay will require students to convince the school board to increase sexual harassment and rape education in schools so that other students do not have to endure the uncertainty that Melinda felt about not knowing what to do or how to heal after being raped. Distribute Writing Assignment #2 to your students and use the rubric attached to Writing Assignment #1 to provide feedback.

WRITING ASSIGNMENT #2 – *Speak*
Writing to persuade

PROMPT
After being raped, Melinda has no idea who to tell or how to deal with what has happened. Instead, she remains silent and tolerates Andy Evans's continuing to sexually harass her at school throughout the year. Your assignment is to write a letter to the school board members in your district, convincing them to increase sexual harassment and rape education in your school.

PREWRITING
Write down all the times Melinda was sexually harassed by Andy at school. Be sure to describe what he was doing to her that made her feel uncomfortable (playing with her hair, sitting too close, etc). Next, reread Melinda's description of being raped and the confusion that followed. Take note of how she felt and reacted immediately following the rape. Then, make a list of all the ways in which the rape has affected Melinda's life.

DRAFTING
Write an introductory paragraph that explains in general why sexual harassment and rape education needs to be increased or created at your school. Mention the current level of sexual harassment and rape education and explain to the school board why this is not sufficient.

In the body paragraphs, outline the specific ways in which being sexually harassed or raped can affect a student. Use your list of ways Andy sexually harassed Melinda to give examples of ways students can be sexually harassed at school. In another paragraph, talk about how being raped or sexually harassed can change the life of that individual, using things that happened to Melinda to guide you. You will also want to discuss the confusion Melinda felt about who to tell and how to deal with being raped.

In your conclusion paragraph, summarize how difficult it can be for a student to be sexually harassed or raped. Point out the current rape or sexual harassment education that takes place at your school now, and end your essay with a powerful reason why the school board needs to increase this type of education.

PROMPT
When you finish the rough draft of your letter, ask a student who sits near you to read it. After reading your rough draft, he/she should tell you what he/she liked best about your work, which parts were difficult to understand, and ways in which your work could be improved. Reread your paper considering your critic's comments, and make the corrections you think are necessary. Ask your classmate what he/she thought of each of the characters/events you chose for your assignment.

PROOFREADING
Do a final proofreading of your paper double-checking your grammar, spelling, organization, and the clarity of your ideas.

LESSON NINE

Objectives
1. To review main ideas, events, and vocabulary of reading Assignment #4
2. To help students recognize similes, metaphors, and personification
3. To help students appreciate why authors use figurative language in their writing

Activity #1
Have students answer the study guide questions for reading Assignment #4 as previously directed.

Activity #2
Review the vocabulary answers from the reading. Make sure students write down the correct answers.

Activity #3
Ask students to come up with reasons why authors use figurative language. Ask them to think about how personification, similes, and metaphors can affect the reader. Spend a few minutes talking about how these literary elements can improve writing and enjoyment of reading.

Next, give students the worksheet on identifying similes, metaphors, and personification in *Speak*. This worksheet will ask them to label examples already provided from the text and add examples of their own. After students have finished the worksheet, go around the room and share the answers to ensure students have correctly labeled the examples. Also, ask students to share the examples they found and check to be sure they are correct.

Key to worksheet:
- Personification: 1, 8, 10
- Simile: 3, 4, 6, 7, 11
- Metaphor: 2, 5, 9, 12

Activity #4
Provide students with markers, colored pencils, or crayons and some blank paper. Instruct them to select an example of figurative language found in *Speak* and illustrate it. They should attempt to draw what the author is saying literally, making the illustrations humorous and showing how figurative language helps readers picture what is going on. Students may use an example from the worksheet or one they found on their own. After students have finished, allow them to share their final product. If possible, display the student drawings in your classroom.

FIGURATIVE LANGUAGE

Authors use figurative language to create a picture in the readers' minds. Most of the time, the author is comparing what is really happening with something people are familiar with, allowing the reader to make a connection with what is happening in the novel. Figurative language also allows the author to express in more a powerful way what is occurring in the novel.

PART ONE: Read the examples of figurative language from *Speak* listed below. Label each example as a simile, metaphor, or personification.

1. Words climb up my throat.

2. I dive into the stream of fourth-period lunch students and swim down the hall to the cafeteria.

3. I have been dropped like a hot Pop Tart on a cold kitchen floor.

4. We are all dressed in down jackets and vests, so we collide and roll like bumper cars at the state fair.

5. There is a beast in my gut, I can hear it scraping away at the inside of my ribs.

6. Her skin is a flat gray color, like underwear washed so many times it's about to fall apart.

7. All the anger whistles out of me like I'm a popped balloon.

8. Lights wink on, the fountains jump, music plays behind the giant ferns, and the mall is open.

9. The card is still there, a white patch of hope with my name on it.

10. I chomp my sandwich and it barfs mustard on my shirt.

11. Her voice sounds like a cold engine that won't turn over.

12. I am a deer frozen in the headlights or a tractor trailer.

PART TWO: Fill in the blanks provided with examples of figurative language you have identified while reading *Speak*.

Simile: _____

Metaphor: _____

Personification: _____

LESSON TEN

Objectives
1. To preview study questions and vocabulary for Assignment #5
2. To begin reading Assignment #5
3. To have students research and read nonfiction related to the book to help connect the book to real life
4. To broaden students' knowledge about topics related to the book

Activity #1
Take students to the library or media center. With students, brainstorm a list on non-fiction topics that could be related to *Speak*. A short list to get you started is included below.
- coping with being raped
- rape among high school students
- sexual harassment among high school students
- sexual harassment/rape education in schools
- crisis centers that can offer victims services
- men who are victims of sexual harassment and rape
- criminal punishment for those who commit the rape
- influences of alcohol on teenagers
- biographies on those who have overcome a problem and have learned to speak up about it to the public
- painters/artists
- the effect of two working parents on the life of a child

Activity #2
Distribute the Nonfiction Assignment Sheet to students. Explain that students should choose a nonfiction topic related to *Speak*. They should read a substantial article related to that topic and complete the Nonfiction Assignment Sheet for that article. Students may use magazines, newspapers, and the Internet as sources.

Activity #3
Bring the class back together and have each student tell what he/she read about.

Note: Compiling the Nonfiction Assignment Sheets into a booklet makes a nice follow-up activity and a handy reference for students.

Activity #4
Look ahead at the study questions and vocabulary for Assignment #5. Tell students that they should read Assignment #5 prior to the next class period. Give them the remainder of this class (if time remains) to complete this assignment.

NONFICTION ASSIGNMENT SHEET
(To be completed after reading the required nonfiction article)

Name _____ Date_____

Title of Nonfiction Read _____

Written By _____ Publication Date _____

I. Factual Summary: Write a short summary of the piece you read.

II. Vocabulary
      1. With which vocabulary words in the piece did you encounter some degree of difficulty?

      2. How did you resolve your lack of understanding with these words?

III. Interpretation: What was the main point the author wanted you to get from reading his work?

IV. Criticism
      1. With which points of the piece did you agree or find easy to accept? Why?

      2. With which points of the piece did you disagree or find difficult to believe? Why?

V. Personal Response: What do you think about this piece? OR How does this piece influence your ideas?

LESSON ELEVEN

<u>Objectives</u>
1. To review main ideas, events, and vocabulary of reading Assignment #5
2. To get students thinking about issues in the news
3. To allow students to express opinions on current issues through writing

Activity #1
Have students answer the study guide questions for reading Assignment #5 as previously directed.

Activity #2
Review the vocabulary answers from the reading. Make sure students write down the correct answers.

Activity #3
Melinda discovers that the bathroom stall doors are covered with student commentary about events and people around school. Melinda uses the bathroom stall door to begin speaking up about Andy Evans by starting her own "thread" called "guys to stay away from." Shortly after she posts this thread, several students respond, carrying on a conversation through writing messages on the bathroom stall door.

Take several pieces of bulletin board paper or butcher paper and place it on a wall. If possible, cut the paper into the shape of bathroom stall doors to add authenticity. You can title the paper "the bathroom stall" or you can leave it blank. Next, tell your students that like Melinda, they will have their own outlet to speak up about issues that concern them. Instruct each student to think about a community, state, national, or international issue they are interested in. Provide students with colored markers and allow them to start a thread on the "stall doors" expressing their opinion or raising awareness about a news issue they feel strongly about. Tell students that this is not a place to gossip about people or school happenings, but a place to speak up about issues that are important to them. Once each student has started a thread, allow for time to read the other topics and add commentary. This assignment should not require any talking and should instead be a place of debate and opinion that takes place on an individual level.

<u>Note</u>: You may want to assign each student a different colored marker to keep track of who is writing what. You may also want to designate the first or last few minutes of class each day for the remainder of the unit for students to continue adding to the discussion.

LESSON TWELVE

<u>Objectives</u>
1. To give students the opportunity to practice writing to inform
2. To improve students' writing ability
3. To connect ideas and facts learned from the book and additional activities with a real-life situation

<u>Activity #1</u>
Distribute the RAFT writing assignment to students. Explain that the purpose of this assignment is to write to inform. Tell students that they should select one of the scenarios listed for their third writing assignment. Explain that the "R" stands for the role they will take, or the point of view they are writing from; the "A" stands for the audience they are writing to; the "F" stands for the format of their writing; and the "T" stands for the topic or task. Quickly go over the different scenarios available to them and give the remaining time for students to complete the assignment.

<u>Note:</u> As students complete this writing assignment, call individuals up for writing conferences on the past two writing assignments. Use the evaluation form to guide you in your conference.

WRITING ASSIGNMENT RAFT
Speak

Directions: Select one of the following writing situations to use as the topic for your essay.

Role *The voice you take on as a writer; this is the perspective you are writing from*	Audience *Who you are writing to; this is the person that will be reading what you write*	Format *The form your writing will take; this is the type of writing you will complete*	Topic/Task *Your purpose for writing; this is the content or reason for your writing assignment*
Melinda	Her parents	Letter	Telling them about what happened the night of the party; why you have been silent; what you have been dealing with
Melinda	Students in a high school	Speech	The dangers of drinking alcohol; the need for communication with parents and other adults
Police Officer	People in the community	Newspaper Article	Warn of punishment connected with the rape/sexual harassment; discuss how much of a problem this is with teenagers
You	Yourself	Journal Entry	Discuss a difficult time you went through; talk about how you dealt with it and how you've learned from the experience and grown as a person

WRITING ASSIGNMENT #3 – *Speak*
Writing to inform

PROMPT
Select one of the scenarios listed on the RAFT writing assignment for the topic of your essay. The role is the point of view you are writing from, the audience is who you are writing to, the format is the type of writing you are doing, and the topic/task is the actual information you are writing about.

PREWRITING
Once you have selected your writing scenario, begin to brainstorm ideas. Remember to think about the role you are writing from and the topic you are writing about. Use your book, notes from the speaker, and notes from the nonfiction articles to help you with your support.

DRAFTING
Write an introductory paragraph that allows the reader to know the role you have assumed and the audience you are writing to. Give a general overview of the points you will make in the body paragraphs of your writing. Use the format of your writing to guide you on how to begin (speech would begin with a little about yourself, letter begins with Dear _____, etc).

In the body paragraphs, give the details of your topic. Use information from the novel, the speaker, and the nonfiction article you read to help provide support. Be sure to reread the topic/task you are writing on and be sure to cover all portions listed there.

In your conclusion paragraph, summarize your main points and conclude the writing assignment. For unity with your writing, you may want to tie in your role and audience once again.

PROMPT
When you finish the rough draft of your composition, ask a student who sits near you to read it. After reading your rough draft, he/she should tell you what he/she liked best about your work, which parts were difficult to understand, and ways in which your work could be improved. Reread your paper considering your critic's comments, and make the corrections you think are necessary. Ask your classmate what he/she thought of each of the characters/events you chose for your assignment.

PROOFREADING
Do a final proofreading of your paper double-checking your grammar, spelling, organization, and the clarity of your ideas.

LESSONS THIRTEEN AND FOURTEEN

Objectives:
1. To allow students to experience the story of *Speak* in a different medium
2. To get students to discuss the similarities and differences between the novel and the movie
3. To appreciate the qualities of both a novel and a movie

Activity #1

Purchase or rent the movie *Speak*. The video is produced by the Lifetime channel and can be purchased via their website or Amazon.com. You may also find the video for rent at Blockbuster or Netflix.

Prepare your students to watch the movie *Speak*. Explain that even though the movie is based on the book, the two are not identical. Have students take notes of the similarities and differences they find while watching the video.

The video will take approximately two days to complete, depending on the length of your class.

Activity #2

Once you have completed the video, hold a class discussion about the similarities and differences between the video and the novel. Talk about the qualities of both movies and novels, discussing the characteristics of each. Be sure to point out the things the producers did well when adapting the book into a movie, and things that can only be conveyed through original text. Allow students to decide which medium they enjoyed best with support for why they felt that way. Also allow students to discuss their feelings towards the actors and actresses selected for each role and how the movie was similar or different from what they have been picturing in their mind throughout the reading.

LESSON FIFTEEN

Objectives
1. To discuss the novel on a deeper than direct-recall level
2. To prepare students for questions and topics covered on the test
3. To allow students to make personal connection with the text

Activity #1
Choose the questions from the Extra Discussion Questions/Writing Assignments which seem most appropriate for your students. A class discussion of these questions is most effective if students have been given the opportunity to formulate answers to the questions prior to the discussion. To this end, you may either have all the students formulate answers to all the questions, divide your class into groups and assign one or more questions to each group, or you could assign one question to each student in your class. The option you choose will make a difference in the amount of class time needed for this activity.

Note: The use of graphic organizers may be helpful to students in preparing their answers. Encourage them to use any diagrams or graphics that they feel are necessary.

EXTRA DISCUSSION QUESTIONS/WRITING ASSIGNMENTS
Speak

Interpretive
1. What are the main conflicts in the story? Describe each fully.

2. What is the setting of the story? How important is the setting to the theme?

3. Describe the author's writing style. Give specific examples to support your answer.

4. List five of Melinda's most important character traits and give examples of each.

5. From what point of view is the story told? Why is that important?

Critical
1. Melinda's friends are ignoring her completely; however, she wants more than anything for them to talk to her again. Explain why Melinda wants her friends to talk to her when they have been treating her so badly.

2. In the chapter called "All Together Now" in reading Assignment #2, Melinda has to choose five verbs and conjugate them. Analyze the five words Melinda chooses and explain how those words are significant in Melinda's life.

3. Describe Melinda's family. Talk about how they interact with each other and what type of role they play in Melinda's life. Determine whether or not the relationship Melinda has with her family is good. Use details from the text to support your answer.

4. In regards to her secret closet at school Melinda says, "I should bribe a janitor to haul all this stuff to my house, make my bedroom more like this, more like home." What does this statement reveal about how Melinda views her home and her bedroom? What conclusions can you make about her life at home versus her life at school?

5. Reread the chapter called "Chat Room" in reading Assignment #5 where students respond to Melinda's warning about Andy Evans on the bathroom wall. What is the student body response like? What does this mean? Why do you think that these girls don't say anything to anyone, but are writing whole paragraphs on the bathroom stall door?

6. As Melinda is healing so is her family. What changes are taking place with her parents and the relationship she has with them?

7. What is the significance of Melinda's trick to eating on the fancy white sofa in the living room?

8. Melinda has chewed on her lips so much that they are scabbed over and hideous-looking. She chews on them all the time, especially to avoid talking. What does this damage to her lips symbolize?

Speak Extra Discussion Questions Page 2

9. Explain why Maya Angelou's books were probably banned from the school library. Why is the Maya Angelou poster in Melinda's closet significant?

10. What does the art project Melinda creates out of left-over turkey bones symbolize?

11. "My parents are arguing. Not a rip-roarer. A simmering argument, a few bubbles splashing on the stove," is an example of what literary element?

12. Melinda studies symbols in her English class. She talks a lot about how her English teacher makes students study Nathaniel Hawthorne and analyze how he uses weather and color to show the moods of characters. In the scene where Melinda is raped, how does the author, Laurie Halse Anderson, use symbols in her writing?

13. The seniors in Melinda's high school get their acceptance/rejection letters from college on April Fool's Day. Of what literary element is this an example?

14. Melinda has to do a report on the suffragettes in her social studies class. Despite her usual lack of effort in class, she writes a phenomenal paper on this topic. Why do you think Melinda admires the suffragettes the way she does?

15. Throughout the book Melinda says things like, "If there is anyone in the entire galaxy I am dying to tell what really happened, it's Rachel" and "She didn't even bother to find out the truth" and "I have worked so hard to forget every second of that stupid party." What literary element do these quotes show?

16. "I stand in the center aisle of the auditorium, a wounded zebra in a *National Geographic* special, looking for someone, anyone, to sit next to" is an example of what literary element?

17. Melinda is silent for most of the book, barely talking to anyone regardless of the situation. Part of her silence is her own decision, while other times she wants to talk but can't. What does her silence symbolize? What changes take place that allow her to begin speaking again?

18. Compare and contrast Melinda with Heather. Why does Melinda hang out with Heather when they are so different?

19. Melinda's parents are very upset with her mid-term grades. They yell at her to get them up. What perceptions do you have about Melinda's parents and their involvement in Melinda's life? Use details from the text to support your answer.

Speak Extra Discussion Questions page 3

20. Explain the symbolism of springtime in relation to Melinda's life.

21. In the yard, Melinda's dad has a tree that is drastically trimmed down because part of it is dead. How is this a metaphor for Melinda's life?

22. Compare and contrast Melinda's performance and attitude in school last year to her performance and attitude in school this year. Use at least five examples from the text to support your answer.

23. At first, Melinda is excited that Andy is flirting with her. She says, "And I thought for just a minute there that I had a boyfriend, I would start high school with a boyfriend, older and stronger and ready to watch out for me." What is ironic about Melinda's image of Andy?

24. "Our frog lies on her back. Waiting for a prince to come and princessify her with a smooch? I stand over her with my knife…My throat closes off. It is hard to breathe. I put out my hand to steady myself against the table. David pins her froggy hands to the dissection tray. He spreads her froggy legs and pins her froggy feet. I have to slice open her belly. She doesn't say a word. She is already dead." How is this frog dissection a metaphor for what happened to Melinda?

25. Mr. Freeman tells Melinda, "This looks like a tree, but it is an average, ordinary, everyday, boring tree. Breathe life into it. Make it bend—trees are flexible, so they don't snap. Scar it, give it a twisted branch—perfect trees don't exist. Nothing is perfect. Flaws are interesting. Be the tree." How is this advice about drawing a tree also a metaphor for Melinda's life?

26. Heather comes to Melinda's house pleading for help and begging to be friends again. What does this act reveal about Heather's character? Compare and contrast how Melinda reacts to Heather now with how she used to react to Heather's pleas when they were friends.

27. Melinda and Rachel had been friends since elementary school. When Melinda opens up to her and reveals she was raped, how would you expect her to react? How does she react instead? Analyze why she reacted this way to such shocking news.

28. Melinda fights back when Andy attempts to rape her again. She holds a glass shard up to his neck and notices, "His lips are paralyzed. He cannot speak." How is this an example of irony?

29. Melinda finally finishes her tree for Mr. Freeman's art class. How was she able to breathe life into her tree? What does her final project say about her life?

Speak Extra Discussion Questions page 4

Critical/Personal Response

1. Melinda refers to Andy as "IT." Do you think this is an appropriate nickname? Explain why you feel this way.

2. Melinda's school is constantly changing their mascot because the school district is nervous about the message it sends to students. The school district won't let the winter assembly have any holiday music at all so as to not discriminate against holiday preferences. The art class has lost its budget and the students will not get new supplies. What do you think of the school district in the book? How realistic is the school district's decision in the book to those in real life? Give examples to support your answer.

3. Using details from the text, compare and contrast David Petrakis with Melinda in terms of academic focus. Which character's academic focus is similar to yours? Explain your answer.

4. When Mr. Freeman offers Melinda a ride home from school she is shocked to hear herself opening up and speaking to him. He tells her, "If you ever need to talk, you know where to find me…You're a good kid. I think you have a lot to say. I'd like to hear it." What is different about Mr. Freeman that allows Melinda to feel comfortable talking to him? How is his approach to talking to Melinda better than that of her parents or other teachers?

5. In the chapter called "Oprah, Sally Jessy, Jerry, and Me" in reading Assignment #4, Melinda imagines what it would be like if her life was discussed on a popular talk show. Analyze the things the three talk show hosts "say" about her being raped. Do you agree with what they have to say? What do you think Melinda should do in this situation?

6. Melinda can't understand why everyone makes a big deal out of her choice to be silent. David tells her, "don't expect to make a difference unless you speak up for yourself." What do you think about Melinda's choice to be silent? Should she speak up about what happened, or is it better to stay silent? Explain your answer.

7. Melinda's dad is having the dead tree in the yard cut down. Melinda describes it as, "Sap oozes from the open sores on the trunk. He is killing the tree. He'll only leave a stump. The tree is dying." In contrast, Melinda's dad tells onlookers, "By cutting off the damage, you make it possible for the tree to grow again. You watch—by the end of the summer, this tree will be the strongest on the block." Later Melinda wonders, "is there a chain saw of the soul, an ax I can take to my memories or fears?" Why do you think Melinda is so upset by the tree being cut down? Why do you think her dad is so optimistic? Which view do you agree with and why?

8. The freshman class of Merryweather High School attends an assembly on the first day of school. As Melinda is listening, she compiles a list entitled "The Top Ten Lies They Tell You In High School." Break down this list into two categories: the items you agree are lies you've heard at your high school and the items you believe are really true at your high school. Next, add two lies of your own and two truths of your own to Melinda's list.

Speak Extra Discussion Questions page 5

9. The author does a good job of describing high school as seen from a teenager's point of view. She pokes fun of teachers, school rules, and other students, just as many teenagers do in real life. Find three passages in this portion of the text you believe to be realistic portrayals of school. Copy down the quote and write a brief description of how that quote is realistic.

Personal Response
1. Melinda says, "It is easier not to say anything. Shut your trap, button your lip, can it. All that crap you hear on TV about communication and expressing feelings is a lie. Nobody really wants to hear what you have to say." Is this true? Explain your answer.

2. Melinda's English class is challenging. She and the other students groan about having to look for literary elements like symbolism while reading a novel. However, as hard as it may be, Melinda starts to get it and says, "Like the whole guilt thing. Of course you know the minister feels guilty and Hester feels guilty, but Nathaniel wants us to know this is a big deal. If he kept repeating, 'She felt guilty, she felt guilty, she felt guilty," it would be a boring book. So he planted symbols, like the weather, and the whole light and dark thing, to show us how poor Hester feels." What do you think of Melinda's statement? Do you enjoy decoding a text and really getting all the meaning from it, or do you prefer just reading it for the plot and missing out on the underlying elements? What about great literary texts that you read it class? Do you enjoy learning about these hidden elements and discussing them? Explain your answer.

3. Melinda's parents get called into school for a meeting with the guidance counselor and the principal. The school has noticed that she skips a lot of school and that she doesn't speak. How do Melinda's parents react? Formulate an alternative way her parents could have handled the situation that may have resulted in a more positive outcome.

4. Reread Melinda's classification of the students at her school at the beginning of the novel. Do you agree that high school students are really divided into cliques the way she describes? Explain your answer.

5. After taking a career quiz, Heather decides that she wants to be a nurse. She then figures out the volunteer work she will do to get experience and what college she wants to go to. Melinda says in response, "How could she know this? I don't know what I'm doing in the next five minutes and she has the text ten years figured out." Which character are you most like in terms of planning for life after high school? Do you plan ahead and have it all figured out like Heather, or are you unsure of what you want to do like Melinda? Explain your answer.

6. The reader finally finds out what really happened at the party and why Melinda called the cops. What are your feelings about what happened? How did you feel when you read what happened?

Speak Extra Discussion Questions page 6

7. Melinda gets in-school suspension (ISS) for skipping another day of school. She says, "The inmates of MISS are commanded to sit and stare at the empty walls. It is supposed to bore us into submission or prepare us for an insane asylum." Do you agree with her description of in-school suspension? Explain why ISS works, or why is doesn't work. Create two alternative punishments that the school could use that would be more effective than in-school suspension.

8. Melinda flips ahead in her Biology book and comments on when students are taught about sex in high school. She says, "We aren't scheduled to learn about that until the eleventh grade." Think about how sexual education is handled in your school. Do you think that it is effective and informative? What could be done to improve it and make it more beneficial to students? Do you agree with the age that sexual education in your school is covered? What age is appropriate to teach about sex in a school setting? Explain why you feel this way.

9. "The climax of mating season is nearly upon us—the Senior Prom. They should cancel school this week. The only things we are learning are who is going with who…who bought a dress in Manhattan, which limo company won't tell if you drink, the most expensive tux place, and on and on and on." Compare and contrast Melinda's description of her school at the time of a big dance and your school when it is time for a big dance.

10. "IT happened. There is no avoiding it, no forgetting. No running away, or flying, or burying, or hiding. Andy Evans raped me in August when I was drunk and too young to know what was happening. It wasn't my fault. He hurt me. It wasn't my fault. And I'm not going to let it kill me. I can grow." At the end of the novel Melinda is able to find her strength and begin the healing process. Each day students are forced to make tough decisions and deal with difficult problems. What realistic alternatives do teenagers have to dealing with serious problems and stress? Formulate at least two alternative choices students can make and explain how those choices are better than the ones Melinda made.

11. The author, Laurie Halse Anderson, had this to say in her author's note:
    When I was a teenager, I couldn't find what I was looking for in contemporary YA fiction. I was searching for characters who struggled with the same problems that haunted me, and nobody seemed to be writing about kids like that…now that I have this amazing opportunity to write books for teenagers, I spend oodles of time listening to them. I've been listening for a couple of years now, trying to figure out what causes them joy and what causes them pain, then looking inside from the stories I can write that might speak to them.

Some people say that in order to really learn something you have to experience it yourself. Others argue that seeing something happening to someone else can help just the same. What do you think? Do you think the author achieved her goal of trying to help teenagers find their voice and inner strength even in the worst circumstances? Do you think the author achieved her goal with you?

Speak Extra Discussion Questions page 6

12. Melinda struggles over whether or not to tell her family and friends what happened. There are several times when she is overcome with a strong desire to talk, but is too nervous about how people will react. Knowing that deep down Melinda really wanted help despite how she hid her problems from her parents, and knowing how much damage keeping her rape experience inside did, how would you handle this problem if one of your friends became withdrawn and quiet? How would you handle this problem if you were a parent and your child was withdrawn? As a parent, what do you think you can do to prevent this from happening to your child?

Quotations

1. "I have entered high school with the wrong hair, the wrong clothes, the wrong attitude. And I don't have anyone to sit with. I am an outcast." Welcome to Merryweather High, RA1

2. "My room belongs to an alien. It's a postcard of who I was in the fifth grade." Home. Work. RA1

3. "This closet is abandon—it has no purpose, no name. It is perfect for me." Burrow, RA1

4. "I am getting better at smiling when people expect it." Acting, RA 1

5. "I want to confess everything, hand over the guilt and mistake and anger to someone else…even if I dump the memory, it will stay with me, staining me." Closet Space, RA2

6. "I bet they'd be divorced right now if I hadn't been born. I'm sure I was a huge disappointment. I'm not pretty or smart or athletic. I'm just like them—an ordinary drone dressed in secrets and lies." Winter Break, RA2

7. "How can I talk to them about that night? How can I start?" Winter Break, RA2

8. "I can smell him over the noise of the metal shop and I drop my poster and the masking tape and I want to throw up and I can smell him and I run and he remembers and he knows. He whispers in my ear." Naming the Monster, RA 2

9. "That's how rabbits survive; they freeze in the presence of predators…BunnyRabbit bolts, leaving fast tracks in the snow. Getaway getaway getaway. Why didn't I run like this before when I was a one-piece talking girl?" Cold Weather and Buses, RA3

10. "I should probably tell someone, just tell someone. Get it over with. Let it out, blurt it out." Escape, RA3

11. "I wonder if Hester tried to say no. She's kind of quiet. We would get along. I can see us, living in the woods, her wearing her A, me with an S maybe, S for silent, for stupid, for scared. S for silly. For shame." Code Breaking, RA3

Speak Extra Discussion Questions page 8

12. "Art is about making mistakes and learning from them." Riding Shotgun, RA3

13. "When people don't express themselves, they die one piece at a time. You'd be shocked how many adults are really dead inside—walking through their days with no idea who they are, just waiting for a heart attack or cancer or a Mack truck to come along and finish the job." Riding Shotgun, RA3

14. "I'm not really here, I'm definitely back at Rachel's crimping my hair and gluing on fake nails, and he smells like beer and mean and he hurts me hurts me hurts me and gets up and zips his jeans and smiles." A Night to Remember, RA3

15. "This looks like a tree, but it is an average, ordinary, everyday, boring tree. Breath life into it. Make it bend—trees are flexible, so they don't snap. Scar it, give it a twisted branch—perfect trees don't exist. Nothing is perfect. Flaws are interesting." Growing Pains, RA4

16. "You can't speak up for your right to be silent. That's letting the bad guys win…Don't expect to make a difference unless you speak up for yourself." Advice from a Smart Mouth, RA4

17. "Did he rape my head, too?" Oprah, Sally Jessy, Jerry, and Me, RA4

18. "He's not chopping it down. He's saving it. Those branches were long dead from disease. All plants are like that. By cutting off the damage, you make it possible for the tree to grow again. You watch—by the end of the summer, this tree will be the strongest on the block." Pruning, RA5

19. "I crouch by the trunk, my fingers stroking the bark, seeking a Braille code, a clue, a message on how to come back to life after my long undersnow dormancy. I have survived. I am here. Confused, screwed up, but here. So, how can I find my way? Is there a chain saw for the soul, an ax I can take to my memories or fears? I dig my fingers into the dirt and squeeze. A small, clean part of me waits to warm and burst through the surface. Some quiet Melindagirl I haven't seen in months. That is the seed I will care for." Pruning, RA5

20. "Sometimes I think high school is one long hazing activity: if you are tough enough to survive this, they'll let you become an adult." Postprom, RA5

21. "His lips are paralyzed. He cannot speak. That's good enough. Me: 'I said no.'" Prey, RA5

22. "My tree is definitely breathing; little shallow breaths like it just shot up through the ground this morning. This one is not perfectly symmetrical. The bark is rough…One of the lower branches is sick. If this tree really lives someplace, that branch better drop soon, so it doesn't kill the whole thing. Roots knob out of the ground and the crown reaches for the sun, tall and healthy. The new growth is the best part." Final Cut, RA5

Speak Extra Discussion Questions page 8

23. "IT happened. There is no avoiding it, no forgetting. No running away, or flying, or burying, or hiding. Andy Evans raped me in August when I was drunk and too young to know what was happening. It wasn't my fault. He hurt me. It wasn't my fault. And I'm not going to let it kill me. I can grow." Final Cut, RA5

24. "The tears dissolve the last block of ice in my throat. I feel the frozen stillness melt down through the inside of me, dripping shards of ice that vanish in a puddle of sunlight on the stained floor. Words float up. Me: 'Let me tell you about it." Final Cut, RA5

LESSON SIXTEEN

Objectives
1. To discuss the novel on a deeper than direct-recall level
2. To prepare students for questions and topics covered on the test
3. To allow students to make personal connection with the text
4. To allow students to practice writing a letter
5. To allow students to show the author their appreciation for the book

Activity #1
Complete the Extra Discussion Questions from the previous assignment.

Activity #2
Give students the Letter to the Author assignment sheet. If needed, review how to write a formal letter with your students. Give them the remainder of the class time to complete the letter. Once you have proofread all the letters and students have had the chance to correct any mistakes, mail the letters to Laurie Halse Anderson at:

    Laurie Halse Anderson
    c/o Penguin Publicity
    345 Hudson Street
    New York, NY 10014

Note: This assignment may be used as an extra credit opportunity. You may offer extra credit for simply writing the letter, or offer extra credit for those who type their letter.

LETTER TO THE AUTHOR

Often times, books are written to make people think about serious issues. Think about the point the author was trying to make in writing this book. Then, compose a letter to the author expressing how this book has affected your life.

Topics to include in your letter:
- What you liked about the book
- How you could relate to this book
- How realistic the book was
- What you learned from the book
- What issues the book made you think about
- How you felt when reading the book
- How you have changed since reading the book
- Anything else you think the author should know

Necessary Elements:
- Your letter must by typed.
- You should begin your letter by saying Dear Mr. (or Ms.) _____,
- You should have an introductory paragraph where you introduce yourself.
- You should have body paragraphs
- You should have a friendly conclusion to the letter.
- Underneath your signature you should include your home address and email address in case the author wishes to write you back.

Remember to proofread this letter and turn it in to me free of errors. I will grade your letter, allow you to make any changes that are needed, and then I will mail your letters to the author of your book. Most authors enjoy receiving letters from readers and like to see how their hard work has affected others. Some authors even respond to letters from their readers, so don't be surprised if you get a reply.

LESSON SEVENTEEN

Objectives:
1. To allow students to present their project to the class
2. To work on students' speaking and presentation skills
3. To generate class discussion about the unit
4. To allow students to connect ideas from the novel with events in their lives and in the lives of those around them
5. To evaluate students' reaction to the unit

Activity #1
Have students present their projects to the class. Ask students to display their project, share the three powerful passages they selected, and share some of their facts they found in their research. If at all possible, allow for discussion of these facts and statistics during presentations. Encourage students to think about the research they did and how it relates to their lives and the lives of other teens.

Activity #2
Give students the *Speak* Unit Reaction sheet. Give them a chance to answer the questions and give feedback about the unit. Read over the feedback and make any necessary adjustments to the unit before you begin to teach it the following year.

SPEAK UNIT REACTION

1. What were your overall impressions of *Speak*?

2. How likely are you to read one of Laurie Halse Anderson's other books?

3. How likely are you to read other young adult books about serious teen issues?

4. What was your favorite assignment in the unit? Why?

5. What was your least favorite assignment in the unit? Why?

6. What was most helpful to you in this unit?

7. What assignment do you think should be changed? Why?

8. What was something that surprised you about your topic while doing research?

9. How did the research and information in the projects presented affect you?

10. How has this book and project affected your life (the decisions you make, how you view others, etc)?

Use the space below to write any other comments you have regarding this unit:

LESSON EIGHTEEN

Objective
To review all of the vocabulary work done in this unit

Activity #1:
Choose one (or more) of the vocabulary review activities listed below and spend your class period as directed in the activity. Some of the materials for these review activities are located in the Vocabulary Resource Materials section in this LitPlan.

VOCABULARY REVIEW ACTIVITIES

1. Divide your class into two teams and have an old-fashioned spelling or definition bee.

2. Give each of your students (or students in groups of two, three or four) a *Speak* Vocabulary Word Search Puzzle. The person (group) to find all of the vocabulary words in the puzzle first wins.

3. Give students a *Speak* Vocabulary Word Search Puzzle without the word list. The person or group to find the most vocabulary words in the puzzle wins.

4. Use a *Speak* Vocabulary Crossword Puzzle. Put the puzzle onto a transparency on the overhead projector (so everyone can see it), and do the puzzle together as a class.

5. Give students a *Speak* Vocabulary Matching Worksheet to do.

6. Divide your class into two teams. Use *Speak* vocabulary words with their letters jumbled as a word list. Student 1 from Team A faces off against Student 1 from Team B. You write the first jumbled word on the board. The first student (1A or 1B) to unscramble the word wins the chance for his/her team to score points. If 1A wins the jumble, go to student 2A and give him/her a definition. He/she must give you the correct spelling of the vocabulary word which fits that definition. If he/she does, Team A scores a point, and you give student 3A a definition for which you expect a correctly spelled matching vocabulary word. Continue giving Team A definitions until some team member makes an incorrect response. An incorrect response sends the game back to the jumbled-word face off, this time with students 2A and 2B. Instead of repeating giving definitions to the first few students of each team, continue with the student after the one who gave the last incorrect response on the team. For example, if Team B wins the jumbled-word face-off, and student 5B gave the last incorrect answer for Team B, you would start this round of definition questions with student 6B, and so on. The team with the most points wins!

7. Have students write a story in which they correctly use as many vocabulary words as possible. Have students read their compositions orally! Post the most original compositions on your bulletin board!

LESSON NINETEEN

Objective:
    To review the main ideas and events in *Speak*

Activity:
Choose one of the review games/activities suggested in this unit and spend your class time as directed there.

REVIEW GAMES/ACTIVITIES *Speak*

1. Ask the class to make up a unit test for *Speak*. The test should have 4 sections: matching, true/false, short answer, and essay. Students may use 1/2 period to make the test and then swap papers and use the other 1/2 class period to take a test a classmate has devised. (open book) You may want to use the unit test included in this packet or take questions from the students' unit tests to formulate your own test.

2. Take 1/2 period for students to make up true and false questions (including the answers). Collect the papers and divide the class into two teams. Draw a big tic-tac-toe board on the chalk board. Make one team X and one team O. Ask questions to each side, giving each student one turn. If the question is answered correctly, that students' team's letter (X or O) is placed in the box. If the answer is incorrect, no letter is placed in the box. The object is to get three in a row like tic-tac-toe. You may want to keep track of the number of games won for each team.

3. Take 1/2 period for students to make up questions (true/false and short answer). Collect the questions. Divide the class into two teams. You'll alternate asking questions to individual members of teams A & B (like in a spelling bee). The question keeps going from A to B until it is correctly answered, then a new question is asked. A correct answer does not allow the team to get another question. Correct answers are +2 points; incorrect answers are -1 point.

4. Have students pair up and quiz each other from their study guides and class notes.

5. Give students a *Speak* crossword puzzle to complete.

6. Play What's My Line?. This is similar to the old television show. Students assume the roles of different characters from the epic. One student gives clues to the class, or to a panel of contestants. The contestants try to guess the identity of the guest. Students may enjoy assisting you in creating rules and procedures for the game.

Review Games Page 2

7. Divide your class into two teams. Use *Speak* crossword words with their letters jumbled as a word list. Student 1 from Team A faces off against Student 1 from Team B. You write the first jumbled word on the board. The first student (1A or 1B) to unscramble the word wins the chance for his/her team to score points. If 1A wins the jumble, go to student 2A and give him/her a clue. He/she must give you the correct word which matches that clue. If he/she does, Team A scores a point, and you give student 3A a clue for which you expect another correct response. Continue giving Team A clues until some team member makes an incorrect response. An incorrect response sends the game back to the jumbled-word face off, this time with students 2A and 2B. Instead of repeating giving clues to the first few students of each team, continue with the student after the one who gave the last incorrect response on the team. For example, if Team B wins the jumbled-word face-off, and student 5B gave the last incorrect answer for Team B, you would start this round of clue questions with student 6B, and so on. The team with the most points wins!

8. Play Jeopardy. Divide the class into two groups. Assign each group a category or book from the epic and have them devise answers for that category. Play the game according to the television show procedures.

9. Play Drawing in the Details. This is similar to Pictionary. Divide students into teams. A student from one team draws a scene from the epic. (You may want to specify the Book or section.) Drawings should be kept simple, to keep the pace lively. Students in the opposing team locate the scene in their books and read it aloud. If they are incorrect, the illustrator's team has a chance to guess. Involve students in setting up a scoring system and any other necessary rules.

10. Take students to a school basketball court. Divide students into two teams. Have student A from team 1 answer a question. If he/she gets it correct, he/she can take a foul shot for an extra point. Next, allow student B from team 2 to answer a question. If the answer is correct he/she can take a foul shot for a bonus point. Continue to ask questions awarding one point for a correct answer and one bonus point for a basket. Point out that part of the reason for the review is based on Melinda's exceptional talent at shooting foul shots, and her need to study to improve her grades.

UNIT TESTS

SHORT ANSWER UNIT TEST 1 - *Speak*

I. Matching/Identify

____ 1. Heather   A. Item Melinda asked her dad to buy for her

____ 2. Efferts   B. The only person Melinda could talk to

____ 3. Rachel   C. What Melinda's dad bought on Thanksgiving

____ 4. MISS   D. Helped Melinda draw a more realistic tree

____ 5. Mr. Freeman   E. Xenophobic and bigoted; unfair to others

____ 6. Andy   F. Gave Melinda a card on Valentine's Day

____ 7. Seeds   G. Inspiration for an art project Melinda did

____ 8. Mr. Neck   H. Exclusive group of girls; performed good deeds around school

____ 9. Bunny   I. Place where Melinda got stuck for skipping class

____ 10. The Marthas   J. Continuously sexually harassed Melinda at school

____ 11. Ivy   K. Where Melinda's mother spent most of her time

____ 12. Sticky Notes   L. Pictured in a painting by the art teacher

____ 13. Donuts   M. How Melinda's family communicated

____ 14. David   N. What Melinda compared herself to

____ 15. Turkey   O. Inspiration for Ivy's art project

____ 16. Art Supplies   P. Found Melinda after Andy tried to rape her a second time

____ 17. Nicole   Q. Doesn't believe that Andy really raped Melinda

____ 18. School Board   R. Biology teacher who creates interesting assignments/labs

____ 19. Ms. Keen   S. Gift Melinda got from her parents

____ 20. Clowns   T. Invited Melinda to a party

Speak Short Answer Unit Test 1 Page 2

II. Short Answer

1. Describe Melinda's hiding place at school.

2. What did Melinda do that has caused everyone in school to hate her?

3. Melinda has chewed on her lips so much that they are scabbed over and hideous looking. She chews on them all the time, especially to avoid talking. What does this damage to her lips symbolize?

4. What upsetting news does Heather give Melinda at lunch one day?

5. Using details from the text, describe Melinda's conflict over whether or not to warn Rachel about Andy.

Speak Short Answer Unit Test 1 Page 3

6. Explain the symbolism of springtime in relation to Melinda's life.

7. What conclusions can you draw based on the response to Melinda's "thread" on the bathroom stall door?

8. Melinda's dad has a tree in the lawn drastically trimmed down since part of it is dead. How is this event a metaphor for Melinda's life?

9. Where does Melinda go on her first bike ride?

10. The first time Melinda was raped she couldn't find her voice and she couldn't find the inner strength to fight back. How does she react when Andy attempts to rape her again?

Speak Short Answer Unit Test 1 Page 4

III. Vocabulary

Write down the vocabulary words. Go back later and write down the correct definition for each word.

1.

2.

3.

4.

5.

6.

7.

8.

9.

10.

Speak Short Answer Unit Test 1 Page 5

IV. Essay
Select *one* of the following topics and respond in an essay:

"My tree is definitely breathing; little shallow breaths like it just shot up through the ground this morning. This one is not perfectly symmetrical. The bark is rough…One of the lower branches is sick. If this tree really lives someplace, that branch better drop soon, so it doesn't kill the whole thing. Roots knob out of the ground and the crown reaches for the sun, tall and healthy. The new growth is the best part." What does Melinda's final project say about her life?

Melinda's English class is challenging. She and the other students groan about having to look for literary elements like symbolism while reading a novel. However, as hard as it may be, Melinda starts to get it and says, "Like the whole guilt thing. Of course you know the minister feels guilty and Hester feels guilty, but Nathaniel wants us to know this is a big deal. If he kept repeating, 'She felt guilty, she felt guilty, she felt guilty," it would be a boring book. So he planted symbols, like the weather, and the whole light and dark thing, to show us how poor Hester feels." What symbols does Laurie Halse Anderson use in *Speak*? Explain at least three symbols used in the book, citing specific examples of each.

SHORT ANSWER UNIT TEST 1 ANSWER KEY – *Speak*

I. Matching/Identify

F	1. Heather	A.	Item Melinda asked her dad to buy for her
K	2. Efferts	B.	The only person Melinda could talk to
Q	3. Rachel	C.	What Melinda's dad bought on Thanksgiving
I	4. MISS	D.	Helped Melinda draw a more realistic tree
B	5. Mr. Freeman	E.	Xenophobic and bigoted; unfair to others
J	6. Andy	F.	Gave Melinda a card on Valentine's Day
A	7. Seeds	G.	Inspiration for an art project Melinda did
E	8. Mr. Neck	H.	Exclusive group of girls; performed good deeds around school
N	9. Bunny	I.	Place where Melinda got stuck for skipping class
H	10. The Marthas	J.	Continuously sexually harassed Melinda at school
D	11. Ivy	K.	Where Melinda's mother spent most of her time
M	12. Sticky Notes	L.	Pictured in a painting by the art teacher
C	13. Donuts	M.	How Melinda's family communicated
T	14. David	N.	What Melinda compared herself to
G	15. Turkey	O.	Inspiration for Ivy's art project
S	16. Art Supplies	P.	Found Melinda after Andy tried to rape her a second time
P	17. Nicole	Q.	Doesn't believe that Andy really raped Melinda
L	18. School Board	R.	Biology teacher who creates interesting assignments/labs
R	19. Ms. Keen	S.	Gift Melinda got from her parents
O	20. Clowns	T.	Invited Melinda to a party

Speak Short Answer Unit Test 1 Answer Key Page 2

II. Short Answer

1. Describe Melinda's hiding place at school.
    Melinda finds an abandoned janitor's closet in the senior wing of the school. The closet smells bad, has dead bugs, and is a little cluttered, but there is an old armchair and a desk. Melinda plans on cleaning up the closet and sneaking in a blanket and some potpourri. She wants to use this closet as a place to hide from other people and skip class.

2. What did Melinda do that has caused everyone in school to hate her?
    Melinda called the cops at Kyle Rodgers's party at the end of the summer.

3. Melinda has chewed on her lips so much that they are scabbed over and hideous looking. She chews on them all the time, especially to avoid talking. What does this damage to her lips symbolize?
    Melinda talks about wanting to swallow herself whole, and having her lips stitched together. All of the references to her lips are a symbol of silence. She chews on them because there is something she wants to say, yet the chewing prevents her from doing it. She has chosen to become silent as much as possible and rarely speaks to anyone, even her teachers and parents. Her damage to her lips symbolizes her damage to her voice in choosing to remain silent. The pain her lips cause her also add to this symbol in that the pain she is feeling is something she can't and won't express.

4. What upsetting news does Heather give Melinda at lunch one day?
    Heather tells Melinda that they are too different and that she doesn't want to be friends with someone who is depressed and weird. She also tells Melinda they can no longer sit together at lunch.

5. Using details from the text, describe Melinda's conflict over whether or not to warn Rachel/Rachelle about Andy.
    Melinda is torn over whether she should warn Rachel/Rachelle about Andy. On one hand, Rachel/Rachelle has been horrible to her all year and deserves whatever happens to her. On the other hand, the two have been friends since a young age, despite what has happened this year. Melinda feels like no one should have to go through what she did, however she knows that getting Rachelle to believe her won't be easy either.

6. Explain the symbolism of springtime in relation to Melinda's life.
    Spring is a season of growth and rebirth. Melinda is suddenly growing from her past and learning to speak up. She starts by standing up to Heather and then feels ready to "arm-wrestle some demons." This is the beginning of a change in Melinda. She is growing and confronting her rape, and discovering a new person inside.

Speak Short Answer Unit Test 1 Answer Key Page 3

7. What conclusions can you draw based on the response to Melinda's "thread" on the bathroom stall door?

    After a short amount of time there are already tons of responses to Melinda's note regarding "Guys to Stay Away From" on the bathroom stall door. Students from the school have responded with other warnings, some even saying that someone should tell the cops about him. This shows that other girls in the school have been victims of Andy Evans as well. One can deduce that others have been raped by him or have had to push him away after quick sexual advances.

8. Melinda's dad has a tree in the lawn drastically trimmed down since part of it is dead. How is this event a metaphor for Melinda's life?

    Melinda's father explains that in order for the tree to continue living it must get rid of the parts that are already dead. He says that this drastic trimming will get rid of all the bad parts and allow the tree to grow back much stronger than it originally was. This is a metaphor in Melinda's life as well. Parts of her are already dead, and in order to move on and not sacrifice her whole life to a few bad parts, she needs to cut out her memories and be able to allow herself to grow once again.

9. Where does Melinda go on her first bike ride?

    Melinda goes to the place where she was raped.

10. The first time Melinda was raped she couldn't find her voice and she couldn't find the inner strength to fight back. How does she react when Andy attempts to rape her again?

    When Andy corners Melinda in the closet and attempts to rape her again she is silent at first. She is terrified and feels like she can't do anything. However, she sees her poster of Maya Angelou and knows that she must do something. She throws small things at him, but he isn't bothered by her feeble attempts to get free. Even though she is pinned down by Andy, she is able to break the mirror with her turkey art project and pick up a shard of glass. Once she has the piece she makes one drop of blood come from Andy's neck and then tells him "No" once again. At that point she is saved by the lacrosse team who heard the struggle and came to see what happened. Melinda was able to fight off Andy on her own and find her strength.

Speak Short Answer Unit Test 1 Answer Key Page 4

III. Vocabulary
  Write down the vocabulary words you have chosen to use for the test.
1.

2.

3.

4.

5.

6.

7.

8.

9.

10.

IV. Essay
Grade the essay according to your own criteria.

"My tree is definitely breathing; little shallow breaths like it just shot up through the ground this morning. This one is not perfectly symmetrical. The bark is rough…One of the lower branches is sick. If this tree really lives someplace, that branch better drop soon, so it doesn't kill the whole thing. Roots knob out of the ground and the crown reaches for the sun, tall and healthy. The new growth is the best part." What does Melinda's final project say about her life?

Melinda's English class is challenging. She and the other students groan about having to look for literary elements like symbolism while reading a novel. However, as hard as it may be, Melinda starts to get it and says, "Like the whole guilt thing. Of course you know the minister feels guilty and Hester feels guilty, but Nathaniel wants us to know this is a big deal. If he kept repeating, 'She felt guilty, she felt guilty, she felt guilty," it would be a boring book. So he planted symbols, like the weather, and the whole light and dark thing, to show us how poor Hester feels." What symbols does Laurie Halse Anderson use in her writing?

SHORT ANSWER UNIT TEST 2 - *Speak*

I. Matching/Identify

____ 1. Rachel   A. Gave Melinda a card on Valentine's Day

____ 2. David   B. Helped Melinda draw a more realistic tree

____ 3. Seeds   C. Found Melinda after Andy tried to rape her a second time

____ 4. Mr. Freeman   D. Biology teacher who creates interesting assignments/labs

____ 5. Ivy   E. Place where Melinda got stuck for skipping class

____ 6. Heather   F. Xenophobic and bigoted; unfair to others

____ 7. The Marthas   G. Exclusive group of girls; performed good deeds around school

____ 8. Turkey   H. How Melinda's family communicated

____ 9. Donuts   I. The only person Melinda could talk to

____ 10. Ms. Keen   J. Inspiration for Ivy's art project

____ 11. Efferts   K. Pictured in a painting by the art teacher

____ 12. Mr. Neck   L. Doesn't believe that Andy really raped Melinda

____ 13. Nicole   M. Inspiration for an art project Melinda did

____ 14. School Board   N. What Melinda's dad bought on Thanksgiving

____ 15. Sticky Notes   O. Gift Melinda got from her parents

____ 16. Andy   P. What Melinda compared herself to

____ 17. MISS   Q. Continuously sexually harassed Melinda at school

____ 18. Art Supplies   R. Item Melinda asked her dad to buy for her

____ 19. Clowns   S. Invited Melinda to a party

____ 20. Bunny   T. Where Melinda's mother spent most of her time

Speak Short Answer Unit Test 2 Page 2

II. Short Answer

1. How does Melinda regard her art class?

2. Describe how the students at Merryweather High School treat Melinda. Give at least three examples from the text to support your answer.

3. Analyze the final version of the art project Melinda makes from left over turkey bones. What does this project symbolize?

4. What is Melinda's reward for going to all her classes for an entire week?

Speak Short Answer Unit Test 2 Page 3

5. Describe Melinda's conflict over the party David Petrakis is throwing at his house to celebrate the basketball victory.

6. What was Melinda's initial reaction to Andy when he approached her at the party?

7. Melinda has to do a report on the suffragettes in her social studies class. Despite her usual lack of effort in class, she writes a phenomenal paper on this topic. Why do you think Melinda admires the suffragettes the way she does?

8. What has happened to the people who were supposed to help Heather decorate for Prom?

9. What happened between Rachel and Andy at the prom?

10. The first time Melinda was raped she couldn't find her voice and she couldn't find the inner strength to fight back. How does she react when Andy attempts to rape her again?

Speak Short Answer Unit Test 2 Page 4

III. Vocabulary

    Write down the vocabulary words. Go back later and write down the correct definitions for the words.

1.

2.

3.

4.

5.

6.

7.

8.

9.

10.

Speak Short Answer Unit Test 2 Page 5

IV. Essay
Select *one* of the following topics and respond in an essay:

"When people don't express themselves, they die one piece at a time. You'd be shocked how many adults are really dead inside—walking through their days with no idea who they are, just waiting for a heart attack or cancer or a Mack truck to come along and finish the job." How does Mr. Freeman's statement apply to Melinda?

Throughout the novel the author makes several references to trees. Aside from trees, springtime and new growth are also used as symbols. Explain how the tree, springtime, and new growth are symbols in Melinda's life.

SHORT ANSWER UNIT TEST 2 ANSWER KEY – *Speak*

I. Matching/Identify

L	1. Rachel	A.	Gave Melinda a card on Valentine's Day
S	2. David	B.	Helped Melinda draw a more realistic tree
R	3. Seeds	C.	Found Melinda after Andy tried to rape her a second time
I	4. Mr. Freeman	D.	Biology teacher who creates interesting assignments/labs
B	5. Ivy	E.	Place where Melinda got stuck for skipping class
A	6. Heather	F.	Xenophobic and bigoted; unfair to others
G	7. The Marthas	G.	Exclusive group of girls; performed good deeds around school
M	8. Turkey	H.	How Melinda's family communicated
N	9. Donuts	I.	The only person Melinda could talk to
D	10. Ms. Keen	J.	Inspiration for Ivy's art project
T	11. Efferts	K.	Pictured in a painting by the art teacher
F	12. Mr. Neck	L.	Doesn't believe that Andy really raped Melinda
C	13. Nicole	M.	Inspiration for an art project Melinda did
K	14. School Board	N.	What Melinda's dad bought on Thanksgiving
H	15. Sticky Notes	O.	Gift Melinda got from her parents
Q	16. Andy	P.	What Melinda compared herself to
E	17. MISS	Q.	Continuously sexually harassed Melinda at school
O	18. Art Supplies	R.	Item Melinda asked her dad to buy for her
J	19. Clowns	S.	Invited Melinda to a party
P	20. Bunny	T.	Where Melinda's mother spent most of her time

Speak Short Answer Unit Test 2 Answer Key Page 2

1. How does Melinda regard her art class?
    The title that comes before her description of art class is "sanctuary." This implies that Melinda feels like her art class is a place she can feel safe. This is the one class during the day that interests her. She actually attempts to do her work and pays attention in class.

2. Describe how the students at Merryweather High School treat Melinda. Give at least three examples from the text to support your answer.
    The students at Melinda's high school all ignore her and glare at her. No one will sit with her in class or at lunch. People bump into her in the hallways and knock her books over. Others openly curse her and tell her how mad they are at her. She is also pushed down three rows of bleachers at the pep rally.

3. Analyze the final version of the art project Melinda makes from left over turkey bones. What does this project symbolize?
    The final version of the art project Melinda makes from left over turkey bones is a museum-like display of the turkey bones, placed together like a skeleton of the dead bird. There is a Barbie doll head on top of the turkey, with tape over her mouth and a fork and knife used as legs. This project represents Melinda as the Barbie head, seemingly perfect at one point, now ripped apart with tape preventing her from speaking. The dead bird skeleton as a body shows how Melinda is dead inside. Mr. Freeman remarks that the project shows pain, and that is exactly what Melinda is trying to express—her pain.

4. What is Melinda's reward for going to all her classes for an entire week?
    Melinda gets to pick out new clothes from the department store her mother works at as a reward for not skipping class. Melinda is happy to get some clothes that fit, but upset they have to come from her mother's store.

5. Describe Melinda's conflict over the party David Petrakis is throwing at his house to celebrate the basketball victory.
    Part of Melinda really wants to go. She sort of likes David and she has no other friends so she is happy to be included in his invitation to a party. However, the other part of her is terrified. She tells herself that he could be lying about his parents being there and it could get out of hand, plus she has a bad track record with parties so far. She ends up telling him no, but is torn the whole way home, part of her wishing she had gone for some fun and the other part hiding in her fear.

Speak Short Answer Unit Test 2 Answer Key Page 3

6. What was Melinda's initial reaction to Andy when he approached her at the party?
   At first, Melinda is very excited that a very cute boy is flirting with her. She wishes that Rachel was there to see this cute guy flirting with her and can't believe how great he looks with his tan and perfectly toned muscles. She begins to think that she will start the school year off with a boyfriend, someone to watch over her and protect her.

7. Melinda has to do a report on the suffragettes in her social studies class. Despite her usual lack of effort in class, she writes a phenomenal paper on this topic. Why do you think Melinda admires the suffragettes the way she does?
   The suffragettes were women who were tired of being mistreated. They felt that women had rights too and they spoke out against the way society treated women. In their time, women could not vote and had no rights. They could not express their opinions or speak out about anything. The suffragettes changed this by fighting for women's rights. They eventually earned women several rights and gave them a voice in society. Melinda admires these women because even though they were mistreated, they still found a way to speak out about what has happening. These were strong women who were tired of feeling the way they did and got together to make changes. Melinda most likely wishes she could be more like these women and find her voice to speak out against her rape.

8. What has happened to the people who were supposed to help Heather decorate for Prom?
   Heather is a part of "the Marthas" and she is supposed to help the girls in the group who are juniors decorate for the Prom. However, the other girls get mono and can't help her decorate. Heather is then desperate to get Melinda to help.

9. What happened between Rachel and Andy at the prom?
   Andy was all over Rachel at the prom while they were dancing. She pushed him off but he wouldn't stop trying to touch her all over. She then confronted him about raping Melinda and stomped off to join her other friends. She almost slapped him, but instead left him humiliated from being dumped by a freshman.

10. The first time Melinda was raped she couldn't find her voice and she couldn't find the inner strength to fight back. How does she react when Andy attempts to rape her again?
    When Andy corners Melinda in the closet and attempts to rape her again she is silent at first. She is terrified and feels like she can't do anything. However, she sees her poster of Maya Angelou and knows that she must do something. She throws small things at him, but he isn't bothered by her feeble attempts to get free. Even though she is pinned down by Andy, she is able to break the mirror with her turkey art project and pick up a shard of glass. Once she has the piece she makes one drop of blood come from Andy's neck and then tells him "No" once again. At that point she is saved by the lacrosse team who heard the struggle and came to see what happened. Melinda was able to fight off Andy on her own and find her strength.

Speak Short Answer Unit Test 2 Answer Key Page 4

III. Vocabulary
    Write down the vocabulary words you have chosen to use.

1.

2.

3.

4.

5.

6.

7.

8.

9.

10.

IV. Essay
Grade the essay according to your own criteria.

"When people don't express themselves, they die one piece at a time. You'd be shocked how many adults are really dead inside—walking through their days with no idea who they are, just waiting for a heart attack or cancer or a Mack truck to come along and finish the job." How does Mr. Freeman's statement apply to Melinda?

Throughout the novel the author makes several references to trees. Aside from trees, springtime and new growth are also used as symbols. Explain how the tree, springtime, and new growth are symbols in Melinda's life.

ADVANCED SHORT ANSWER UNIT TEST *Speak*

I. Matching/Identify

____ 1. Rachel   A. Gave Melinda a card on Valentine's Day

____ 2. David   B. Helped Melinda draw a more realistic tree

____ 3. Seeds   C. Found Melinda after Andy tried to rape her a second time

____ 4. Mr. Freeman   D. Biology teacher who creates interesting assignments/labs

____ 5. Ivy   E. Place where Melinda got stuck for skipping class

____ 6. Heather   F. Xenophobic and bigoted; unfair to others

____ 7. The Marthas   G. Exclusive group of girls; performed good deeds around school

____ 8. Turkey   H. How Melinda's family communicated

____ 9. Donuts   I. The only person Melinda could talk to

____ 10. Ms. Keen   J. Inspiration for Ivy's art project

____ 11. Efferts   K. Pictured in a painting by the art teacher

____ 12. Mr. Neck   L. Doesn't believe that Andy really raped Melinda

____ 13. Nicole   M. Inspiration for an art project Melinda did

____ 14. School Board   N. What Melinda's dad bought on Thanksgiving

____ 15. Sticky Notes   O. Gift Melinda got from her parents

____ 16. Andy   P. What Melinda compared herself to

____ 17. MISS   Q. Continuously sexually harassed Melinda at school

____ 18. Art Supplies   R. Item Melinda asked her dad to buy for her

____ 19. Clowns   S. Invited Melinda to a party

____ 20. Bunny   T. Where Melinda's mother spent most of her time

Speak Advanced Short Answer Unit Test Page 2

II. Short Answer

1. Describe Melinda's family. Talk about how they interact with each other and what type of role they play in Melinda's life. Determine whether or not the relationship Melinda has with her family is good. Use details from the text to support your answer.

2. Melinda's writes a warning about Andy Evans on the bathroom stall. What is the student body response like? What does this mean? Why do you think that these girls don't say anything to anyone, but are writing whole paragraphs on the bathroom stall door?

3. Melinda has chewed on her lips so much that they are scabbed over and hideous looking. She chews on them all the time, especially to avoid talking. What does this damage to her lips symbolize?

Speak Advanced Short Answer Unit Test Page 3

4. "Our frog lies on her back. Waiting for a prince to come and princessify her with a smooch? I stand over her with my knife…My throat closes off. It is hard to breathe. I put out my hand to steady myself against the table. David pins her froggy hands to the dissection tray. He spreads her froggy legs and pins her froggy feet. I have to slice open her belly. She doesn't say a word. She is already dead." How is this frog dissection a metaphor for what happened to Melinda?

5. Mr. Freeman tells Melinda, "This looks like a tree, but it is an average, ordinary, everyday, boring tree. Breathe life into it. Make it bend—trees are flexible, so they don't snap. Scar it, give it a twisted branch—perfect trees don't exist. Nothing is perfect. Flaws are interesting. Be the tree." How is this advice about drawing a tree also a metaphor for Melinda's life?

6. Heather comes to Melinda's house pleading for help and begging to be friends again. What does this act reveal about Heather's character? Compare and contrast how Melinda reacts to Heather now, with how she used to react to Heather's pleas when they were friends.

Speak Advanced Short Answer Unit Test Page 4

7. Melinda's English class is challenging. She and the other students groan about having to look for literary elements like symbolism while reading a novel. However, as hard as it may be, Melinda starts to get it and says, "Like the whole guilt thing. Of course you know the minister feels guilty and Hester feels guilty, but Nathaniel wants us to know this is a big deal. If he kept repeating, 'She felt guilty, she felt guilty, she felt guilty," it would be a boring book. So he planted symbols, like the weather, and the whole light and dark thing, to show us how poor Hester feels." What symbols does Laurie Halse Anderson use throughout *Speak*?

Speak Advanced Short Answer Unit Test Page 5

III. Vocabulary

    Write down the vocabulary words given, then write a paragraph or two about *Speak* correctly using all of the words.

Speak Advanced Short Answer Unit Test Page 6

IV. Essay

"I wonder if Hester [from *The Scarlet Letter*] tried to say no. She's kind of quiet. We would get along. I can see us, living in the woods, her wearing her A, me with an S maybe, S for silent, for stupid, for scared. S for silly. For shame." Melinda mentions five "S" words to describe how she feels after being raped. For each word, determine whether or not Melinda is justified in feeling this way. If so, explain why that feeling is valid. If not, explain why Melinda should not be carrying that emotion around inside of her.

MULTIPLE CHOICE UNIT TEST 1 - *Speak*

I. Matching/Identify

____ 1. Heather   A. Item Melinda asked her dad to buy for her

____ 2. Efferts   B. The only person Melinda could talk to

____ 3. Rachel   C. What Melinda's dad bought on Thanksgiving

____ 4. MISS   D. Helped Melinda draw a more realistic tree

____ 5. Mr. Freeman   E. Xenophobic and bigoted; unfair to others

____ 6. Andy   F. Gave Melinda a card on Valentine's Day

____ 7. Seeds   G. Inspiration for an art project Melinda did

____ 8. Mr. Neck   H. Exclusive group of girls; performed good deeds around school

____ 9. Bunny   I. Place where Melinda got stuck for skipping class

____ 10. The Marthas   J. Continuously sexually harassed Melinda at school

____ 11. Ivy   K. Where Melinda's mother spent most of her time

____ 12. Sticky Notes   L. Pictured in a painting by the art teacher

____ 13. Donuts   M. How Melinda's family communicated

____ 14. David   N. What Melinda compared herself to

____ 15. Turkey   O. Inspiration for Ivy's art project

____ 16. Art Supplies   P. Found Melinda after Andy tried to rape her a second time

____ 17. Nicole   Q. Doesn't believe that Andy really raped Melinda

____ 18. School Board   R. Biology teacher who creates interesting assignments/labs

____ 19. Ms. Keen   S. Gift Melinda got from her parents

____ 20. Clowns   T. Invited Melinda to a party

Speak Multiple Choice Unit Test 1 Page 2

II. Multiple Choice

1. Where was Melinda's hiding place at school?
    A. In an abandoned janitor's closet
    B. Under the bleachers in the gym
    C. In the first stall of the girl's bathroom
    D. In the back of the library where few ever go

2. What did Melinda do that has caused everyone in school to hate her?
    A. Complained to the principal and got the school mascot changed
    B. Let the school down in the state tennis tournament, causing the school to lose to their rival
    C. Called the cops at a huge party over the summer
    D. Told the principal that the most popular boy at school had drugs, which led to his arrest

3. Melinda has chewed on her lips so much that they are scabbed over and hideous-looking. She chewed on them a lot, especially to avoid talking. What does the damage to her lips symbolize?
    A. Her rising anger towards her parents and everyone at school
    B. Her nervousness about school because when she thinks about her low grades she gets nervous about getting in trouble with her parents
    C. Her constant need for attention from her family and friends–she feels ignored and people act concerned when they see her damaged lips
    D. Her lack of ability to speak up about what has happened to her and the silence that has taken over instead

4. What upsetting news did Heather give Melinda at lunch one day?
    A. Heather told Melinda she was moving to New Jersey with her mom since her parents were getting a divorce.
    B. Heather told Melinda she was dating her ex-boyfriend.
    C. Heather told Melinda they could no longer be friends or sit together at lunch.
    D. Heather told Melinda she had been diagnosed with cancer and will have to start treatment next week.

Speak Multiple Choice Unit Test 1 Page 3

5. Melinda is unsure of whether or not she should warn Rachel about Andy. Which of the following is NOT a reason she is nervous to tell her?
    A. Melinda has seen Andy act like a perfect gentleman with his last few girlfriends. Melinda thinks that maybe Andy learned his lesson and has changed, giving her no reason to warn Rachel.
    B. Rachel has been really mean and hurtful to Melinda all year long. Melinda is unsure of whether or not she should be nice to her or let her learn her own lesson.
    C. Melinda is worried that Rachel won't believe her since they haven't been friends for so long. She also thinks Rachel will accuse her of being jealous.
    D. Melinda is too nervous to tell someone how she knows Andy is bad. She feels embarrassed that other people might find out what happened to her.

6. How is the new spring season a symbol for what is taking place in Melinda's life?
    A. Spring is a symbol of growth and rebirth. After standing up to Heather, Melinda realizes she is ready to start wrestling with her problems. She is changing and growing as a person and is ready to come alive again after feeling dead for so long.
    B. Spring is a symbol for bright colors and cheer. Melinda has begun wearing brightly colored clothes to school with the new season. She has also started spreading cheer by becoming active in the school. She is now going to all her classes, and even helps make others happy by decorating the school for Prom.
    C. Spring is a symbol for new life. Melinda finds out that her mom and dad are expecting a new baby, which will bring new life into their home. Melinda also sees birds laying eggs, bees pollinating flowers, and new life springing up all around her.
    D. Spring is a symbol for devotion to Christianity. With the Easter holiday, Melinda realizes that she has been neglecting her faith in God. She renews her faith and begins to rely on God to help her get past the horrible memories of being raped.

7. How does the reader know that other people in the school have been sexually harassed or raped by Andy Evans?
    A. Rachel calls the cops after Andy almost rapes her. When the police come to the school, several other girls come forward and confess that Andy did the same thing to them.
    B. Students in Melinda's English class have to write an essay on the worst event they have ever experienced. Several of the girls in the class write their essay about being sexually harassed or raped by Andy Evans.
    C. Ms. Keen overhears Andy talking with his friends in the hall. He is bragging about all the girls he has slept with and comments that the girls don't have a choice. The teacher reports it and everyone in the school finds out.
    D. Melinda writes on the bathroom stall to stay away from Andy. After only a few days, tons of other students have responded to her comment on the stall. Several students write about their experience with Andy, alluding to other cases of sexual harassment and rape.

Speak Multiple Choice Unit Test 1 Page 4

8. Melinda's dad has a tree in the lawn drastically trimmed down since part of it is dead. How is this event a metaphor for Melinda's life?
    A. Melinda's dad feels that if something is damaged it needs to be gotten rid of. He says that the dead parts are of no use to the tree, so they should be cut off. This implies that Melinda needs to get rid of every memory of the rape, and completely get rid of her dead parts.
    B. Melinda's dad explains that the dead parts of the tree must be cut off so they don't take over and kill the tree. He says once the dead parts are gone, the tree will grow back stronger. This implies that Melinda needs to cut the dead parts out of her life in order to grow into a stronger individual and heal.
    C. Melinda's dad thinks that the dead branches are ugly. He feels they are making their tree look bad, and the yard as well. Melinda feels like her personality has been ugly ever since the rape. She knows that is has effected her whole life and that she needs to get rid of the ugly thoughts and feelings to not make the rest of her life look bad.
    D. Melinda's dad feels like cutting the dead branches of the tree off is part of the spring cleaning process. This implies that Melinda needs to do a "spring cleaning" of her life and get rid of the problems she is having. She needs to clean up her life and get back on track.

9. Where does Melinda go on her first bike ride?
    A. Melinda goes to the park to find inspiration for her tree project.
    B. Melinda goes to Heather's house to try and become friends again.
    C. Melinda goes to Andy's house to confront him about the rape.
    D. Melinda goes to the spot she was raped.

10. The first time Melinda was raped she couldn't find her voice and she couldn't find the inner strength to fight back. How does she react when Andy attempts to rape her again?
    A. She is unable to move or fight back. She is silent and terrified that it is happening again, but luckily the lacrosse team is walking by the closet. They are curious what all the noise is and when they see what is happening they stop it and run to get help.
    B. She is frozen at first but then starts to scream "no". Andy covers her mouth and tries to overpower her, but Melinda uses all her strength to break a mirror and hold a shard of glass to his throat.
    C. She immediately begins to fight him off. She has thought about this moment of revenge for so long that she knows exactly what to do. Andy ends up regretting that he ever messed with Melinda.
    D. She is unable to move or fight back. All she can do is cry while Andy rapes her again. Afterwards, she is angry at herself for letting it happen again and vows to tell the cops and get Andy arrested for raping her.

Speak Multiple Choice Unit Test 1 Page 5

III. Vocabulary - Match the correct definitions to the words.

____ 1. indoctrination   A. choke; suffocate; smother

____ 2. inconspicuous   B. a remote control mechanism

____ 3. errant   C. not noticeable; invisible

____ 4. pseudo   D. a difficult problem; a dilemma

____ 5. floundering   E. the inner sense of what is right or wrong

____ 6. degrading   F. humiliating; disgrace; dishonor

____ 7. xenophobic   G. to act clumsily or in confusion

____ 8. drone   H. poorly adjusted in one's social circumstances

____ 9. demure   I. something or someone that is inspiring to an artist

____ 10. conundrum   J. shy, modest, coy

____ 11. asphyxiated   K. persistent; stubborn; vicious; not easily pulled apart

____ 12. reluctance   L. unwillingness; resisting

____ 13. tenacious   M. teaching someone to accept an idea without criticism

____ 14. momentum   N. to cut off, clear, or remove

____ 15. conscience   O. straying from the right course

____ 16. muse   P. unreasonable fear or hatred of foreigners

____ 17. coaxes   Q. force or speed of movement; motion

____ 18. maladjusted   R. something that has been described but not proven

____ 19. pruning   S. pretend; fake; false

____ 20. allegedly   T. to persuade by pleading or flattery

Speak Multiple Choice Unit Test 1 Page 6

IV. Essay

"My tree is definitely breathing; little shallow breaths like it just shot up through the ground this morning. This one is not perfectly symmetrical. The bark is rough…One of the lower branches is sick. If this tree really lives someplace, that branch better drop soon, so it doesn't kill the whole thing. Roots knob out of the ground and the crown reaches for the sun, tall and healthy. The new growth is the best part." What does Melinda's final project say about her life? Write a complete essay and use information from the book to support your points.

MULTIPLE CHOICE UNIT TEST 2 – *Speak*

I. Matching/Identify

____ 1. Rachel   A. Gave Melinda a card on Valentine's Day

____ 2. David   B. Helped Melinda draw a more realistic tree

____ 3. Seeds   C. Found Melinda after Andy tried to rape her a second time

____ 4. Mr. Freeman   D. Biology teacher who creates interesting assignments/labs

____ 5. Ivy   E. Place where Melinda got stuck for skipping class

____ 6. Heather   F. Xenophobic and bigoted; unfair to others

____ 7. The Marthas   G. Exclusive group of girls; performed good deeds around school

____ 8. Turkey   H. How Melinda's family communicated

____ 9. Donuts   I. The only person Melinda could talk to

____ 10. Ms. Keen   J. Inspiration for Ivy's art project

____ 11. Efferts   K. Pictured in a painting by the art teacher

____ 12. Mr. Neck   L. Doesn't believe that Andy really raped Melinda

____ 13. Nicole   M. Inspiration for an art project Melinda did

____ 14. School Board   N. What Melinda's dad bought on Thanksgiving

____ 15. Sticky Notes   O. Gift Melinda got from her parents

____ 16. Andy   P. What Melinda compared herself to

____ 17. MISS   Q. Continuously sexually harassed Melinda at school

____ 18. Art Supplies   R. Item Melinda asked her dad to buy for her

____ 19. Clowns   S. Invited Melinda to a party

____ 20. Bunny   T. Where Melinda's mother spent most of her time

Speak Multiple Choice Unit Test 2 Page 2

II. Multiple Choice

1. How does Melinda regard her art class?
    - A. She thinks it is a joke and the assignments are stupid. She thinks Mr. Freeman is crazy.
    - B. She feels safe in the art room. This is the only class where she pays attention and does her work.
    - C. She thinks it is boring. She skips this class as much as possible.
    - D. She thinks it is scary since Mr. Freeman is mean and yells at the class.

2. Describe how the students at Merryweather High School treat Melinda.
    - A. She is a fairly popular girl. There are several guys interested in dating her and she has lots of friends.
    - B. She is hated by everyone. They push her down the bleachers, knock her books over, and say rude things to her in the halls.
    - C. She is viewed as a really caring person. She volunteers to throw a teacher appreciation party, tutors other students after school, and is involved in several clubs.
    - D. She is viewed as an athletic girl. She is the star of the tennis team and runs cross country.

3. Melinda used leftover turkey bones to make skeleton of a turkey. She then places a Barbie doll head on top of the turkey, placing a piece of tape over the mouth. What does the final version of this project symbolize?
    - A. The project expresses the pain Melinda is going through.
    - B. The project expresses the commercialism of Thanksgiving.
    - C. The project expresses a lost childhood.
    - D. The project expresses the mistreatment of animals that are butchered for food.

4. What is Melinda's reward for going to all her classes for an entire week?
    - A. She gets to go to a party at David Petrakis' house.
    - B. She gets to buy new clothes at the department store where her mother works.
    - C. She gets to buy new art supplies.
    - D. She gets to decide where her family goes for dinner to celebrate.

5. Why is Melinda hesitant to go to the party David Petrakis is throwing at his house to celebrate the basketball victory?
    - A. She is scared she won't know very many people and won't have anyone to talk to.
    - B. She is nervous that if she goes David will know she likes him. She has never been kissed before and worries David might try to kiss her if she goes.
    - C. She is scared his parents might not be there and the party will get out of hand.
    - D. She has a big project due in her social studies class. If she doesn't do a good job on it then she will fail the semester and have to go to summer school.

Speak Multiple Choice Unit Test 2 Page 3

6. What was Melinda's initial reaction to Andy when he approached her at the party last summer?
    A. She thought he was ugly and felt like she could find someone better to date. Plus, she already had a crush on someone else.
    B. She was really excited because someone so cute was flirting with her. She thought she would begin high school with a boyfriend to watch over her.
    C. She was worried because she knew her friend Rachel liked him. She didn't want Rachel to be mad that Andy was flirting with her instead.
    D. She was relieved to finally see someone she knew at the party. Even though they weren't good friends, she was happy to have someone to talk to.

7. Melinda has to do a report on the suffragettes in her social studies class. Despite her usual lack of effort in class, she writes a phenomenal paper on this topic. Why do you think Melinda admires the suffragettes the way she does?
    A. Melinda found out that her great-great grandmother was part of the suffragettes and wants to do a good job on her paper to honor her memory.
    B. The suffragettes are women who speak up for women's rights. Melinda admires their ability to voice their opinions and fight for their rights.
    C. A scholarship is available to all students who write a paper on why they admire the suffragettes. Knowing that her parents don't have much money, Melinda takes an interest in writing about the suffragettes and in hopes of getting the scholarship.
    D. The suffragettes are women who are active in government. Melinda has been active in her school government all year and admires them because they have the same goals and interests.

8. What has happened to the people who were supposed to help Heather decorate for Prom?
    A. All the other girls who are supposed to help get mono and are home sick.
    B. The sophomores and juniors decide that since Heather is only a freshman they will make her do all the work.
    C. All the other girls got caught skipping school and going to the mall instead. They are all suspended and are not allowed to help.
    D. All the girls are fighting over how to decorate for Prom. No one can agree so they all go home mad, leaving Heather to do it all.

Speak Multiple Choice Unit Test 2 Page 4

9. What happened between Rachel and Andy at the prom?
    A. Andy decides it is too embarrassing for him to date a freshman. He breaks up with Rachel and she causes a big scene in front of everyone.
    B. Rachel overhears Andy talking to his friends about later that night. She finds out he got a hotel room and was expecting more than she was ready to give. She sneaks off and leaves Andy at the Prom with no explanation about where she went.
    C. Andy was all over Rachel at the Prom. She kept pushing him off but he wouldn't stop. She ends up dumping him and leaving him alone at the Prom.
    D. Andy's ex-girlfriend shows up at the Prom alone. He dances with her while Rachel is in the bathroom. Rachel comes back to see them dancing and gets angry.

10. The first time Melinda was raped she couldn't find her voice and she couldn't find the inner strength to fight back. How does she react when Andy attempts to rape her again?
    A. She is unable to move or fight back. She is silent and terrified that it is happening again, but luckily the lacrosse team is walking by the closet. They are curious what all the noise is and when they see what is happening they stop it and run to get help.
    B. She is frozen at first but then starts to scream "No". Andy covers her mouth and tries to overpower her, but Melinda uses all her strength to break a mirror and hold a shard of glass to his throat.
    C. She immediately begins to fight him off. She has thought about this moment of revenge for so long that she knows exactly what to do. Andy ends up regretting that he ever messed with Melinda.
    D. She is unable to move or fight back. All she can do is cry while Andy rapes her again. Afterwards, she is angry at herself for letting it happen again and vows to tell the cops and get Andy arrested for raping her.

III. Essay
Select *one* of the following topics and respond in an essay:

"When people don't express themselves, they die one piece at a time. You'd be shocked how many adults are really dead inside—walking through their days with no idea who they are, just waiting for a heart attack or cancer or a Mack truck to come along and finish the job." How does Mr. Freeman's statement apply to Melinda?

Throughout the novel the author makes several references to trees. Aside from trees, springtime and new growth are also used as symbols. Explain how the tree, springtime, and new growth are symbols in Melinda's life.

Speak Multiple Choice Unit Test 2 Page 5

IV. Vocabulary - Match the correct definitions to the words.

____ 1. demerit   A. keeping the same behavior, form, pattern, or principles

____ 2. sanctuary   B. wrongful; illegal; failure to fulfill a duty or obligation

____ 3. wan   C. a sacred place offering refuge or safety

____ 4. pseudo   D. to withdraw or go back

____ 5. burrow   E. pensive; thoughtful in a sad way; longing yearning

____ 6. refurbished   F. persistent; stubborn; vicious; not easily pulled apart

____ 7. retreat   G. intolerant of any other beliefs or opinions

____ 8. sensibilities   H. a hole or hideout animals use to take shelter; a hideout

____ 9. dormant   I. a person who is rejected; an outcast

____ 10. imbeciles   J. to abuse newcomers with humiliating tricks and ridicule

____ 11. dynamics   K. a group of stupid or silly people

____ 12. bigoted   L. the social, intellectual, or physical forces that characterize a system or group

____ 13. wistful   M. a mark against someone for misconduct

____ 14. consistency   N. emotions or feelings

____ 15. hazing   O. pretend; fake; false

____ 16. tenacious   P. to persuade by pleading or flattery

____ 17. delinquency   Q. not clear or definite; hazy

____ 18. coaxes   R. dark; gloomy; pale in color; sickly; unhappy

____ 19. vaguely   S. to make clean, bright, or fresh again; renovate

____ 20. leper   T. inactive; lying asleep; not erupting

ANSWER SHEET - *Speak*
Multiple Choice Unit Test 1

	Matching	Mult Choice	Vocabulary
1	F	A	M
2	K	C	C
3	Q	D	O
4	I	C	S
5	B	A	G
6	J	A	F
7	A	D	P
8	E	B	B
9	N	D	J
10	H	B	D
11	D		A
12	M		L
13	C		K
14	T		Q
15	G		E
16	S		I
17	P		T
18	L		H
19	R		N
20	O		R

ANSWER SHEET - *Speak*
Multiple Choice Unit Test 2

	Matching	Mult Choice	Vocabulary
1	L	B	M
2	S	B	C
3	R	A	R
4	I	B	O
5	B	C	H
6	A	B	S
7	G	B	D
8	M	A	N
9	N	C	T
10	D	B	K
11	T		L
12	F		G
13	C		E
14	K		A
15	H		J
16	Q		F
17	E		B
18	O		P
19	J		Q
20	P		I

UNIT RESOURCE MATERIALS

BULLETIN BOARD IDEAS - *Speak*

1. Save one corner of the board for the best of students' *Speak* writing assignments.

2. Take one of the word search puzzles from the extra activities packet and with a marker copy it over in a large size on the bulletin board. Write the clue words to find to one side. Invite students prior to and after class to find the words and circle them on the bulletin board.

3. Write several of the most significant quotations from the book onto the board on brightly colored paper.

4. Make a bulletin board listing the vocabulary words for this unit. As you complete sections of the novel and discuss the vocabulary for each section, write the definitions on the bulletin board. (If your board is one students face frequently, it will help them learn the words.)

5. Post photos and information about people who had found their voice and spoke up about something that has happened to them. Examples would be students who travel the country and talk to high school about drinking after being in an accident, etc.

6. Make your bulletin board look like a refrigerator. Put a stack of sticky notes near the bulletin board and invite students to post messages to each other before and after class, communicating like Melinda and her parents do.

7. Decorate the bulletin board with a large tree. Put powerful quotes from the novel on leaves or other parts of the tree. Note: If you don't have room to put your KWL chart on a wall, you can place your class KWL chart on your bulletin board.

8. Make a bulletin board covering other teen issues. You could post facts/statistics on drinking, drugs, dropping out, pregnancy, etc. Next to each group of statistics, print out colorful copies of young adult novel covers that deal with those issues. A list of suggested novels and the topics they cover is listed below.
   - Drugs: *Crank* by Ellen Hopkins, *Go Ask Alice* by an anonymous author, *Impulse* by Ellen Hopkins
   - Cutting: *Cut* by Patricia McCormick
   - Eating Disorders: *Diary of an Anorexic Girl* by Morgan Menzie
   - Teen Pregnancy: *Make Lemonade* by Virginia Euwer Wolff, *Catalyst* by Laurie Halse Anderson
   - Dropping Out: *Make Lemonade* by Virginia Euwer Wolff
   - Suicide: *Burn Journals* by Brent Runyon, *Impulse* by Ellen Hopkins
   - Violence against others: *Burned* by Ellen Hopkins
   - Rape: *Lovely Bones* by Alice Sebold

Bulletin Board Ideas Continued *Speak*

9. Make a bulletin board with colorful copies of Laurie Halse Anderson's other novels. Write a short tease for the book to get students interested.

10. Post artwork done by some of the artists mentioned in *Speak* (Picasso, Kahlo, Monet, O'Keeffe, Pollock, Dali). Make an area that allows students to post their own artwork that they feel expresses their thoughts or feelings.

EXTRA ACTIVITIES - *Speak*

One of the difficulties in teaching a novel is that all students don't read at the same speed. One student who likes to read may take the book home and finish it in a day or two. Sometimes a few students finish the in-class assignments early. The problem, then, is finding suitable extra activities for students.

One thing that seems to help is to keep a little library in the classroom. For this unit on *Speak*, you might check out from the school library *Catalyst, Prom, Twisted,* or *Fever 1793,* other novels by Laurie Halse Anderson. The novels listed in the bulletin board section that cover other teen issues would also be good titles to have in your room as well. Bring in articles or books about sexual harassment, rape, crisis centers, art, art history, local art galleries or shows, safe dating, care of trees, nervous habits, outward signs of inward trouble, decorating for special events, or careers in teaching, counseling, art, or law enforcement.

Other things you may keep on hand are puzzles. We have made some relating directly to *Speak* for you. Feel free to duplicate them for your students to use.

Some students may like to draw. You might devise a contest or allow some extra-credit grade for students who draw characters or scenes from *Speak.* Note, too, that if the students do not want to keep their drawings you may pick up some extra bulletin board materials this way. If you have a contest and you supply the prize (a free iTunes download or something like that perhaps), you could, possibly, make the drawing itself a non-returnable entry fee.

The pages which follow contain games, puzzles and worksheets. The keys, when appropriate, immediately follow the puzzle or worksheet. There are two main groups of activities: one group for the unit; that is, generally relating to *Speak* text, and another group of activities related strictly to *Speak* vocabulary.

Directions for these games, puzzles and worksheets are self-explanatory. The object here is to provide you with extra materials you may use in any way you choose.

MORE ACTIVITIES - *Speak*

1. Have students work together to make a time line chronology of the events in the story. Take a large piece of construction paper and on one wall (or however you can physically arrange it in your room) make the events of the story along it. Students may want to add drawings or cut-out pictures to represent the events (as well as a written statement).

2. Have students design a book cover (front and back and inside flaps) for *Speak*.

3. Have students design a bulletin board (ready to be put up; not just sketched) for *Speak*.

4. Have students group the chapters together to show the larger structure of the novel. Have them explain why they chose the divisions they made.

5. Have students choose one chapter of the book (with sufficient dialogue) to rewrite as a play. In conjunction with this assignment, have students write a composition explaining the difficulties they encountered in changing from one written form to another.

6. Have students write out the characters in the book and cast famous actors and actress for a movie version of the novel. Instruct students to write a brief explanation as to why the actor/actress they selected would be perfect for the part.

7. Read the list of topics Melinda and her classmates write about in their English class (Word Work, RA 2). Have students select one or more of the topics she had to write about and write their own essay on those topics.

8. In the Platinum Edition of *Speak* there is an interview with Laurie Halse Anderson. Have students read the interview and discuss or write a short response.

9. Allow students to explore Laurie Halse Anderson's web page (www.writerlady.com).

10. Have students write a short sequel, discussing how Melinda changes in her sophomore year of high school.

11. Have students find a poem that has the same topic/theme as *Speak*. Tell students to put the text of the poem on a paper and make a collage of magazine word/picture cutouts to decorate the area around the poem.

12. Have students select a character from the book and complete the "I Am" poem from that character's point of view. (see handout on following page)

13. Have students respond to a quote when Melinda talks about the world being perfect, just for an instant. (see handout on following page)

"I Am" Poem

Complete this "I am" poem. You may select any character from the book to do this poem about. Be sure to write from his or her point of view and think about the things he or she would feel. You may use some short one word answers, but do not make each line only a few words. You should try to provide support from the novel to really develop this poem so that it reveals information and insight about the character you select.

I am (2 characteristics your character has)
I wonder (something your character wonders)
I hear (something real or imaginary your character hears)
I see (something real or imaginary your character sees)
I want (something your character desires)
I am (the first line of the poem repeated)

I pretend (something your character pretends to do)
I feel (something real or imaginary your character feels emotionally)
I touch (something real or imaginary your character would touch physically)
I worry (something your characters worries about)
I cry (something that makes your character upset)
I am (the first line of the poem repeated)

I understand (something your character knows)
I say (something your character believes in)
I dream (something your character would dream about)
I try (something your character makes an effort to do)
I hope (something your character hopes for)
I am (the first line of the poem repeated)

Literature and Composition Activity: *Speak*

Directions: Read the passage below. Then, on your own sheet of paper, answer question one in a well-developed paragraph. After you have answered the first question, reread the passage and answer the second question in a well-developed paragraph. Once you finish, you should have written two detailed paragraphs fully answering both questions.

Excerpt from *Speak* by Laurie Halse Anderson

"Applesmell soaks the air. One time when I was little, my parents took me to an orchard. Daddy set me high in an apple tree. It was like falling up into a storybook, yummy and red and leaf and the branch not shaking a bit. Bees bumbled through the air, so stuffed with apple they couldn't be bothered to sting me. The sun warmed my hair, and a wind pushed my mother into my father's arms, and all the apple-picking parents and children smiled for a long, long minute."
-Page 66

1. Melinda, the main character in *Speak*, remembers a time in her childhood where the whole world seemed perfect, just for a moment. Think about your most treasured childhood memories. Select a time in your childhood that seemed perfect and write a detailed paragraph describing that moment.

2. Melinda's parents took her to the apple orchard in order for her to experience something they felt was important in life. Think about all the wonderful things to experience in this world. Then, in a well-developed paragraph, explain what one experience you will want to show your child when they are young. Explain specifically what you want your child to experience and why.

Speak Word List

No.	Word	Clue/Definition
1.	ANDY	Melinda warned others about him.
2.	ANGELOU	Her books were banned in the library; her poster picture was in Melinda's closet
3.	ART	Kind of supplies Melinda's parents gave her for Christmas
4.	BIKE	Melinda's transportation to the place where she was raped
5.	BONES	Turkey ___ were part of Melinda's art project
6.	BUNNY	Animal Melinda compared herself to when she was near Andy
7.	CLOSET	Place where Melinda hides at school
8.	CLOWNS	Inspiration for Ivy's art project
9.	COPS	People Melinda called after being raped
10.	DAVID	He invited Melinda to a party after a basketball game.
11.	DONUTS	What Melinda's dad bought on Thanksgiving
12.	DRACULA	Book Melinda read on Halloween
13.	EFFERTS	Store where Melinda's mother worked
14.	FREEMAN	Only teacher Melinda talked to
15.	FRIENDS	Heather told Melinda they could no longer be this.
16.	FROG	Subject of dissection in Biology
17.	GLASS	Melinda held this up to Andy's throat when he tried to rape her again
18.	GLOBE	Mr. Freeman used this to assign art projects at the beginning of the year.
19.	HEATHER	She gave Melinda a card on Valentine's Day.
20.	IT	Nickname Melinda had for Andy
21.	IVY	She helped Melinda draw a more realistic tree.
22.	KEEN	Biology teacher who created interesting assignments
23.	LAB	David was Melinda's ___ partner in Biology class.
24.	LIBRARY	Place where Melinda confided in Rachel about being raped
25.	LIPS	They were scabbed over from being chewed on
26.	MARTHAS	Exclusive group of girls who performed good deeds
27.	MASCOT	The school ___ kept changing
28.	MELINDA	She found her voice in the end.
29.	MERRYWEATHER	Name of the school
30.	MISS	Place where Melinda got stuck for skipping class
31.	MODEL	Part-time job for Heather
32.	NECK	Teacher who was xenophobic, bigoted, and unjust
33.	NECKLACE	Heather's Christmas present from Melinda
34.	NICOLE	She found Melinda after Andy tried to rape her a second time.
35.	NOTES	Method of communication for Melinda's family
36.	PICASSO	Artist who inspired Melinda
37.	PROM	Event at which Rachel broke up with Andy
38.	RACHEL	Melinda tried to warn her about Andy.
39.	SEEDS	Melinda asked her dad to buy these for her.
40.	SPEAK	Melinda did not do this very much.
41.	STALL	Melinda wrote a warning about Andy on the bathroom ___ door.
42.	SUFFRAGETTES	Women Melinda did a report about
43.	TAXI	Melinda's dad's transportation to the airport
44.	TENNIS	Melinda almost beat Nicole at this sport
45.	TREE	Major symbol in the book; subject of Melinda's art project

WORD SEARCH - Speak

```
A V F C B U N N Y H M P P N T Y Y X M P
L N T L L L Y L X E J I P V L J O B B
D E G A V O W G H A R V S C I K C D D L
S R F E X Z S L C T R C Y O A J D S E P
N D A F L I N E Z H Y W H L R S A V L M
R N K C E O P Y T E W K R E P H S L P P
X A K Z U R U H F R E K K R T R G O S M
D P C M C L T T Z Y A J F R T T O M U Z
X A N H D C A S R G T H A R V T K M F N
F R V X E Q R F B X H M G P Y F T G F N
D S T I J L H D R D E T B C S R J S R S
O W L B D V Y E Z I R E M L D E N R A X
N B T T N R M C M J E N X O H E S F G K
U I F N A G A A G I M N K W J M E B E S
T K E R H A S L L M S I D N L A T B T S
S E B C N E C K A R T S F S E N O B T W
K I M D O F O C S I B R P D P L N A E M
L F Y Y L P T E S D O I E E G V L L S H
J R P Y R J S N F G L J Z E A L T K K P
K G C F F M R B K W H N P S X K T S N R
```

ANDY	DRACULA	LAB	NOTES
ANGELOU	EFFERTS	LIBRARY	PICASSO
ART	FREEMAN	LIPS	PROM
BIKE	FRIENDS	MARTHAS	RACHEL
BONES	FROG	MASCOT	SEEDS
BUNNY	GLASS	MERRYWEATHER	SPEAK
CLOSET	GLOBE	MISS	STALL
CLOWNS	HEATHER	MODEL	SUFFRAGETTES
COPS	IT	NECK	TAXI
DAVID	IVY	NECKLACE	TENNIS
DONUTS	KEEN	NICOLE	TREE

WORD SEARCH ANSWER KEY - Speak

ANDY	DRACULA	LAB	NOTES
ANGELOU	EFFERTS	LIBRARY	PICASSO
ART	FREEMAN	LIPS	PROM
BIKE	FRIENDS	MARTHAS	RACHEL
BONES	FROG	MASCOT	SEEDS
BUNNY	GLASS	MERRYWEATHER	SPEAK
CLOSET	GLOBE	MISS	STALL
CLOWNS	HEATHER	MODEL	SUFFRAGETTES
COPS	IT	NECK	TAXI
DAVID	IVY	NECKLACE	TENNIS
DONUTS	KEEN	NICOLE	TREE

CROSSWORD - Speak

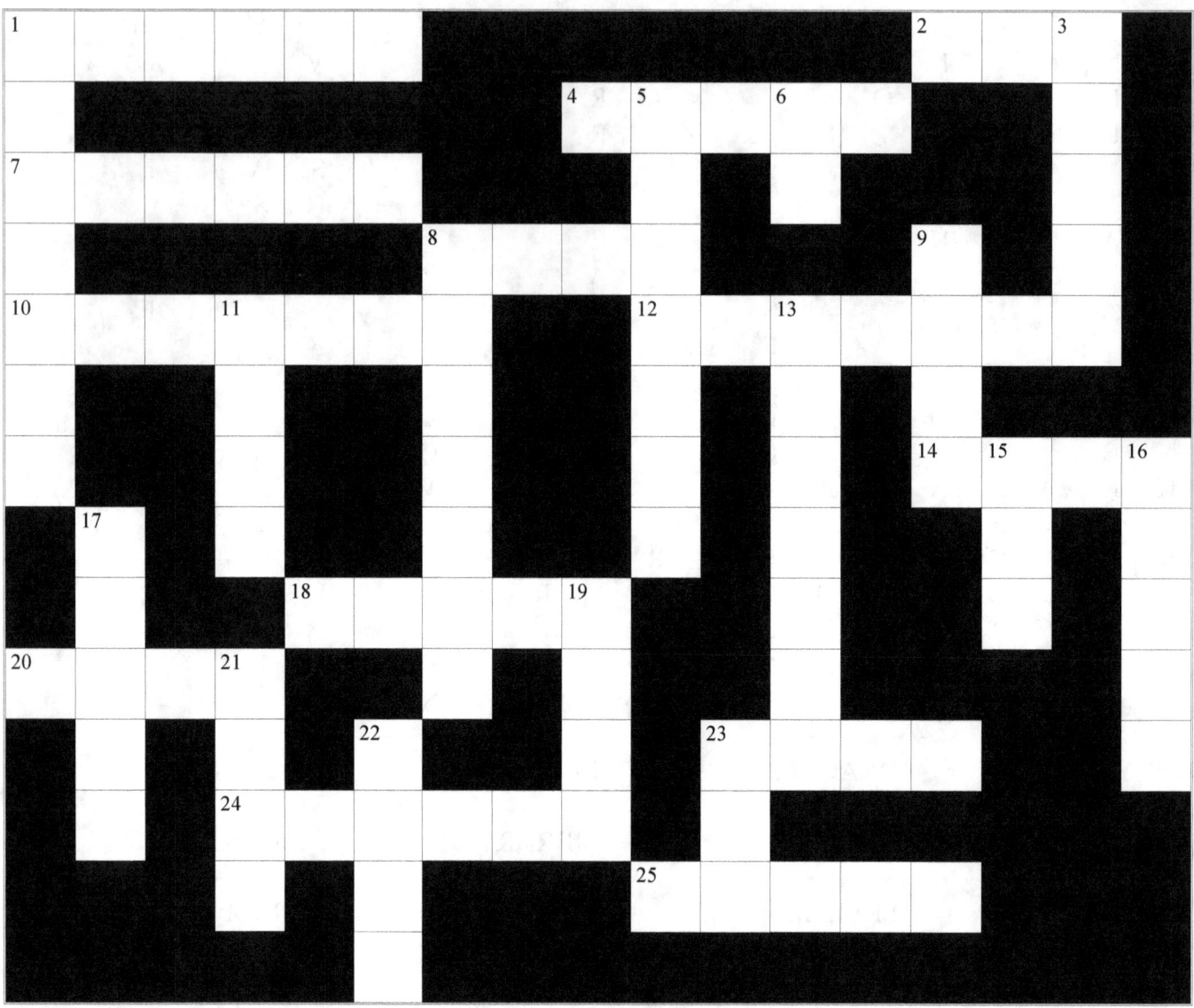

Across
1. The school ___ kept changing
2. David was Melinda's ___ partner in Biology class.
4. He invited Melinda to a party after a basketball game.
7. Melinda tried to warn her about Andy.
8. Subject of dissection in Biology
10. She gave Melinda a card on Valentine's Day.
12. Store where Melinda's mother worked
14. Place where Melinda got stuck for skipping class
18. Part-time job for Heather
20. Biology teacher who created interesting assignments
23. Melinda warned others about him.
24. Inspiration for Ivy's art project
25. Melinda wrote a warning about Andy on the bathroom ___ door.

Down
1. Exclusive group of girls who performed good deeds
3. Turkey ___ were part of Melinda's art project
5. Her books were banned in the library; her poster picture was in Melinda's closet
6. Nickname Melinda had for Andy
8. Heather told Melinda they could no longer be this.
9. Event at which Rachel broke up with Andy
11. Melinda's dad's transportation to the airport
13. Only teacher Melinda talked to
15. She helped Melinda draw a more realistic tree.
16. Melinda asked her dad to buy these for her.
17. Melinda did not do this very much.
19. They were scabbed over from being chewed on
21. Teacher who was xenophobic, bigoted, and unjust
22. People Melinda called after being raped
23. Kind of supplies Melinda's parents gave her for Christmas

CROSSWORD ANSWER KEY - Speak

	1 M	A	S	C	O	T					2 L	A	3 B
	A					4 D	5 A	V	6 I	D			O
	7 R	A	C	H	E	L		N		T			N
	T				8 F	R	O	G			9 P		E
10 H	E	11 A	T	H	E	R		12 E	13 F	F	E	R	T S
A		X			I			L		R		O	
S					E			O		E	14 M	15 I	16 S
	17 S	I			N			U		E		V	E
	P		18 M	O	D	E	19 L			M		Y	E
20 K	E	21 N			S		I			A			D
	A		22 C				P		23 A	N	D	Y	S
	K	24 C	L	O	W	N	S		R				
			K	P				25 S	T	A	L	L	
				S									

Across
1. The school ___ kept changing
2. David was Melinda's ___ partner in Biology class.
4. He invited Melinda to a party after a basketball game.
7. Melinda tried to warn her about Andy.
8. Subject of dissection in Biology
10. She gave Melinda a card on Valentine's Day.
12. Store where Melinda's mother worked
14. Place where Melinda got stuck for skipping class
18. Part-time job for Heather
20. Biology teacher who created interesting assignments
23. Melinda warned others about him.
24. Inspiration for Ivy's art project
25. Melinda wrote a warning about Andy on the bathroom ___ door.

Down
1. Exclusive group of girls who performed good deeds
3. Turkey ___ were part of Melinda's art project
5. Her books were banned in the library; her poster picture was in Melinda's closet
6. Nickname Melinda had for Andy
8. Heather told Melinda they could no longer be this.
9. Event at which Rachel broke up with Andy
11. Melinda's dad's transportation to the airport
13. Only teacher Melinda talked to
15. She helped Melinda draw a more realistic tree.
16. Melinda asked her dad to buy these for her.
17. Melinda did not do this very much.
19. They were scabbed over from being chewed on
21. Teacher who was xenophobic, bigoted, and unjust
22. People Melinda called after being raped
14923. Kind of supplies Melinda's parents gave her for Christmas

MATCHING 1 - Speak

___ 1. NICOLE   A. She found Melinda after Andy tried to rape her a second time.
___ 2. PICASSO   B. Heather told Melinda they could no longer be this.
___ 3. TAXI   C. Subject of dissection in Biology
___ 4. BONES   D. Method of communication for Melinda's family
___ 5. MERRYWEATHER   E. Melinda wrote a warning about Andy on the bathroom ___ door.
___ 6. LIBRARY   F. Inspiration for Ivy's art project
___ 7. FREEMAN   G. She found her voice in the end.
___ 8. MELINDA   H. Her books were banned in the library; her poster picture was in Melinda's closet
___ 9. NOTES   I. Only teacher Melinda talked to
___ 10. DAVID   J. The school ___ kept changing
___ 11. GLOBE   K. Artist who inspired Melinda
___ 12. FROG   L. Name of the school
___ 13. TREE   M. Women Melinda did a report about
___ 14. GLASS   N. Melinda tried to warn her about Andy.
___ 15. IT   O. Turkey ___ were part of Melinda's art project
___ 16. STALL   P. Place where Melinda confided in Rachel about being raped
___ 17. LIPS   Q. He invited Melinda to a party after a basketball game.
___ 18. MASCOT   R. Melinda's transportation to the place where she was raped
___ 19. SUFFRAGETTES S. Melinda's dad's transportation to the airport
___ 20. BIKE   T. Heather's Christmas present from Melinda
___ 21. CLOWNS   U. Melinda held this up to Andy's throat when he tried to rape her again
___ 22. RACHEL   V. Mr. Freeman used this to assign art projects at the beginning of the year.
___ 23. FRIENDS   W. They were scabbed over from being chewed on
___ 24. ANGELOU   X. Nickname Melinda had for Andy
___ 25. NECKLACE   Y. Major symbol in the book; subject of Melinda's art project

MATCHING 1 ANSWER KEY - Speak

A - 1. NICOLE	A.	She found Melinda after Andy tried to rape her a second time.
K - 2. PICASSO	B.	Heather told Melinda they could no longer be this.
S - 3. TAXI	C.	Subject of dissection in Biology
O - 4. BONES	D.	Method of communication for Melinda's family
L - 5. MERRYWEATHER	E.	Melinda wrote a warning about Andy on the bathroom ___ door.
P - 6. LIBRARY	F.	Inspiration for Ivy's art project
I - 7. FREEMAN	G.	She found her voice in the end.
G - 8. MELINDA	H.	Her books were banned in the library; her poster picture was in Melinda's closet
D - 9. NOTES	I.	Only teacher Melinda talked to
Q - 10. DAVID	J.	The school ___ kept changing
V - 11. GLOBE	K.	Artist who inspired Melinda
C - 12. FROG	L.	Name of the school
Y - 13. TREE	M.	Women Melinda did a report about
U - 14. GLASS	N.	Melinda tried to warn her about Andy.
X - 15. IT	O.	Turkey ___ were part of Melinda's art project
E - 16. STALL	P.	Place where Melinda confided in Rachel about being raped
W 17. LIPS	Q.	He invited Melinda to a party after a basketball game.
J - 18. MASCOT	R.	Melinda's transportation to the place where she was raped
M - 19. SUFFRAGETTES	S.	Melinda's dad's transportation to the airport
R - 20. BIKE	T.	Heather's Christmas present from Melinda
F - 21. CLOWNS	U.	Melinda held this up to Andy's throat when he tried to rape her again
N - 22. RACHEL	V.	Mr. Freeman used this to assign art projects at the beginning of the year.
B - 23. FRIENDS	W.	They were scabbed over from being chewed on
H - 24. ANGELOU	X.	Nickname Melinda had for Andy
T - 25. NECKLACE	Y.	Major symbol in the book; subject of Melinda's art project

MATCHING 2 - Speak

___ 1. PROM A. Melinda warned others about him.
___ 2. MASCOT B. Women Melinda did a report about
___ 3. NOTES C. They were scabbed over from being chewed on
___ 4. DONUTS D. The school ___ kept changing
___ 5. FROG E. Only teacher Melinda talked to
___ 6. GLOBE F. She found Melinda after Andy tried to rape her a second time.
___ 7. KEEN G. Event at which Rachel broke up with Andy
___ 8. DAVID H. Melinda's dad's transportation to the airport
___ 9. CLOWNS I. Heather's Christmas present from Melinda
___ 10. STALL J. Subject of dissection in Biology
___ 11. NICOLE K. Method of communication for Melinda's family
___ 12. FRIENDS L. Mr. Freeman used this to assign art projects at the beginning of the year.
___ 13. NECKLACE M. Nickname Melinda had for Andy
___ 14. PICASSO N. Major symbol in the book; subject of Melinda's art project
___ 15. LIPS O. He invited Melinda to a party after a basketball game.
___ 16. COPS P. Exclusive group of girls who performed good deeds
___ 17. TREE Q. Biology teacher who created interesting assignments
___ 18. ANDY R. Teacher who was xenophobic, bigoted, and unjust
___ 19. EFFERTS S. Heather told Melinda they could no longer be this.
___ 20. TAXI T. Artist who inspired Melinda
___ 21. NECK U. Melinda wrote a warning about Andy on the bathroom ___ door.
___ 22. IT V. What Melinda's dad bought on Thanksgiving
___ 23. SUFFRAGETTES W. Store where Melinda's mother worked
___ 24. FREEMAN X. People Melinda called after being raped
___ 25. MARTHAS Y. Inspiration for Ivy's art project

MATCHING 2 ANSWER KEY - Speak

G - 1.	PROM	A. Melinda warned others about him.
D - 2.	MASCOT	B. Women Melinda did a report about
K - 3.	NOTES	C. They were scabbed over from being chewed on
V - 4.	DONUTS	D. The school ___ kept changing
J - 5.	FROG	E. Only teacher Melinda talked to
L - 6.	GLOBE	F. She found Melinda after Andy tried to rape her a second time.
Q - 7.	KEEN	G. Event at which Rachel broke up with Andy
O - 8.	DAVID	H. Melinda's dad's transportation to the airport
Y - 9.	CLOWNS	I. Heather's Christmas present from Melinda
U -10.	STALL	J. Subject of dissection in Biology
F -11.	NICOLE	K. Method of communication for Melinda's family
S -12.	FRIENDS	L. Mr. Freeman used this to assign art projects at the beginning of the year.
I -13.	NECKLACE	M. Nickname Melinda had for Andy
T -14.	PICASSO	N. Major symbol in the book; subject of Melinda's art project
C -15.	LIPS	O. He invited Melinda to a party after a basketball game.
X -16.	COPS	P. Exclusive group of girls who performed good deeds
N -17.	TREE	Q. Biology teacher who created interesting assignments
A -18.	ANDY	R. Teacher who was xenophobic, bigoted, and unjust
W 19.	EFFERTS	S. Heather told Melinda they could no longer be this.
H -20.	TAXI	T. Artist who inspired Melinda
R -21.	NECK	U. Melinda wrote a warning about Andy on the bathroom ___ door.
M -22.	IT	V. What Melinda's dad bought on Thanksgiving
B -23.	SUFFRAGETTES	W. Store where Melinda's mother worked
E -24.	FREEMAN	X. People Melinda called after being raped
P -25.	MARTHAS	Y. Inspiration for Ivy's art project

JUGGLE LETTERS 1 - Speak

1. IENSDFR = 1. _____
   Heather told Melinda they could no longer be this.

2. ASLTL = 2. _____
   Melinda wrote a warning about Andy on the bathroom ___ door.

3. RLBIAYR = 3. _____
   Place where Melinda confided in Rachel about being raped

4. IADVD = 4. _____
   He invited Melinda to a party after a basketball game.

5. SSGLA = 5. _____
   Melinda held this up to Andy's throat when he tried to rape her again

6. TRA = 6. _____
   Kind of supplies Melinda's parents gave her for Christmas

7. ORFG = 7. _____
   Subject of dissection in Biology

8. ERFAEUSFTGTS = 8. _____
   Women Melinda did a report about

9. MEFANRE = 9. _____
   Only teacher Melinda talked to

10. RLUAACD =10. _____
    Book Melinda read on Halloween

11. ATMHSRA =11. _____
    Exclusive group of girls who performed good deeds

12. PSAKE =12. _____
    Melinda did not do this very much.

13. ETRE =13. _____
    Major symbol in the book; subject of Melinda's art project

14. OILCEN =14. _____
    She found Melinda after Andy tried to rape her a second time.

15. EKEN =15. _____
    Biology teacher who created interesting assignments

JUGGLE LETTERS 1 ANSWER KEY - Speak

1. IENSDFR = 1. FRIENDS
Heather told Melinda they could no longer be this.

2. ASLTL = 2. STALL
Melinda wrote a warning about Andy on the bathroom ___ door.

3. RLBIAYR = 3. LIBRARY
Place where Melinda confided in Rachel about being raped

4. IADVD = 4. DAVID
He invited Melinda to a party after a basketball game.

5. SSGLA = 5. GLASS
Melinda held this up to Andy's throat when he tried to rape her again

6. TRA = 6. ART
Kind of supplies Melinda's parents gave her for Christmas

7. ORFG = 7. FROG
Subject of dissection in Biology

8. ERFAEUSFTGTS = 8. SUFFRAGETTES
Women Melinda did a report about

9. MEFANRE = 9. FREEMAN
Only teacher Melinda talked to

10. RLUAACD = 10. DRACULA
Book Melinda read on Halloween

11. ATMHSRA = 11. MARTHAS
Exclusive group of girls who performed good deeds

12. PSAKE = 12. SPEAK
Melinda did not do this very much.

13. ETRE = 13. TREE
Major symbol in the book; subject of Melinda's art project

14. OILCEN = 14. NICOLE
She found Melinda after Andy tried to rape her a second time.

15. EKEN = 15. KEEN
Biology teacher who created interesting assignments

JUGGLE LETTERS 2 - Speak

1. OPRM = 1. _____
   Event at which Rachel broke up with Andy

2. EESRTATFFSGU = 2. _____
   Women Melinda did a report about

3. YIARBRL = 3. _____
   Place where Melinda confided in Rachel about being raped

4. LAUEOGN = 4. _____
   Her books were banned in the library; her poster picture was in Melinda's closet

5. NMAIELD = 5. _____
   She found her voice in the end.

6. ULADCAR = 6. _____
   Book Melinda read on Halloween

7. KNEE = 7. _____
   Biology teacher who created interesting assignments

8. ISSM = 8. _____
   Place where Melinda got stuck for skipping class

9. UNNYB = 9. _____
   Animal Melinda compared herself to when she was near Andy

10. ESERFFT =10. _____
    Store where Melinda's mother worked

11. UTNDOS =11. _____
    What Melinda's dad bought on Thanksgiving

12. DNYA =12. _____
    Melinda warned others about him.

13. OLWSNC =13. _____
    Inspiration for Ivy's art project

14. YAETRHWREEMR =14. _____
    Name of the school

15. EMEFNAR =15. _____
    Only teacher Melinda talked to

JUGGLE LETTERS 2 ANSWER KEY - Speak

1. OPRM = 1. PROM
   Event at which Rachel broke up with Andy

2. EESRTATFFSGU = 2. SUFFRAGETTES
   Women Melinda did a report about

3. YIARBRL = 3. LIBRARY
   Place where Melinda confided in Rachel about being raped

4. LAUEOGN = 4. ANGELOU
   Her books were banned in the library; her poster picture was in Melinda's closet

5. NMAIELD = 5. MELINDA
   She found her voice in the end.

6. ULADCAR = 6. DRACULA
   Book Melinda read on Halloween

7. KNEE = 7. KEEN
   Biology teacher who created interesting assignments

8. ISSM = 8. MISS
   Place where Melinda got stuck for skipping class

9. UNNYB = 9. BUNNY
   Animal Melinda compared herself to when she was near Andy

10. ESERFFT = 10. EFFERTS
    Store where Melinda's mother worked

11. UTNDOS = 11. DONUTS
    What Melinda's dad bought on Thanksgiving

12. DNYA = 12. ANDY
    Melinda warned others about him.

13. OLWSNC = 13. CLOWNS
    Inspiration for Ivy's art project

14. YAETRHWREEMR = 14. MERRYWEATHER
    Name of the school

15. EMEFNAR = 15. FREEMAN
    Only teacher Melinda talked to

# VOCABULARY RESOURCE MATERIALS

Speak Vocabulary Word List

No.	Word	Clue/Definition
1.	ALLEGEDLY	Supposedly; put forth as true but not proven
2.	ASPHYXIATED	Suffocated; smothered; choked
3.	BANISHED	Forced to leave
4.	BIGOTED	Intolerant of any other beliefs or opinions
5.	BURROW	Hole or hideout animals use to take shelter
6.	COAXES	Persuades by pleading or flattery
7.	CONSCIENCE	Inner sense of what is right or wrong
8.	CONSISTENCY	Keeping the same behavior, form, pattern, or principles
9.	CONUNDRUM	Difficult problem; dilemma
10.	DEGRADING	Humiliating; disgracing
11.	DELINQUENCY	Failure to fulfill a duty or obligation; something wrongful or illegal
12.	DEMERIT	Mark against someone for misconduct
13.	DEMURE	Shy; modest; coy
14.	DENSE	Dull or slow-witted
15.	DEVIOUS	Deceitful; not straightforward
16.	DORMANT	Inactive; asleep
17.	DRONE	Remote-controlled mechanism
18.	DYNAMICS	Social, intellectual, or physical forces that characterize a system or group
19.	ERRANT	Straying from the right course
20.	FLOUNDERING	Acting clumsily or in confusion
21.	FOSTER	Something that nourishes or cares for
22.	GENETICS	Science of heredity and genes
23.	HAZING	Abusing newcomers with humiliating tricks and ridicule before they become a part of the group
24.	IMBECILES	Stupid or silly people
25.	INCITING	Stirring up (trouble); egging-on
26.	INCONSPICUOUS	Not noticeable
27.	INCRIMINATE	Make someone appear guilty of a crime
28.	INDOCTRINATION	Teaching someone to accept an idea or principle without criticism
29.	LEPER	Outcast
30.	MALADJUSTED	Not in sync with one's circumstances
31.	MOMENTUM	Force or speed of movement
32.	MUSE	Something or someone that is inspiring to an artist
33.	PRUNING	Cutting; clipping
34.	PSEUDO	Fake; false; pretend
35.	RECESSIVE	Going to the back; a gene that does not produce
36.	REFURBISHED	Made clean, bright, or fresh again
37.	RELUCTANCE	Unwillingness; resistance
38.	REPUTATION	How the public views or regards an individual
39.	RETREAT	Withdraw; go back
40.	REVOLUTIONARY	Supporting radical change or innovation
41.	SANCTUARY	Sacred place offering refuge or safety
42.	SENSIBILITIES	Emotions; feelings
43.	SUBJECTIVITY	Based on personal feelings rather than facts
44.	SUBMISSION	Surrendering power to another
45.	TENACIOUS	Persistent; stubborn; won't give up
46.	VAGUELY	Not clear or definite
47.	VESPIARY	Nest of social wasps
48.	WAN	Pale in color; sickly-looking
49.	WISTFUL	Thoughtful in a sad way; longing
50.	XENOPHOBIC	Having an unreasonable fear or hatred of foreigners

VOCABULARY WORD SEARCH - Speak

```
I V S J S W Y C N E U Q N I L E D W M H
N A E T A N I M I R C N I L E P E R A J
C S N S G Y V S E Y E R F H X N V Z L N
I P S V P W M T T R E C Q B O D I G A S
T H I A B I S M G F E Y E R S N O B D F
I Y B G N O A F U T U P D S G V U Y J M
N X I U F C N R N A E L U Y S Y S J U B
G I L E M T B A Y E K N B T N I H M S T
E A I L P I R S J R S I A I A A V D T P
N T T Y S R E U D T U P M C G T M E E W
E E I H E B C B Q E B O U B I O I I D F
T D E Y U P N J C R M B S Y E O T O C G
I D S R D T A E O E I U E L P C U E N S
C X R A O I T C N C S D R D R C I S D X
S O Z U R R C T U O S O H E U P G L K V
W B Q T C E U I N A I R S G N H L Q E M
J Y L C S M L V D X O M L E I G T J T S
T J W N K E E I R E N A D L N R C K V J
N H E A P D R T U S J N W L G K T C F Z
J D V S W W N Y M G B T B A N I S H E D
```

ALLEGEDLY	DEVIOUS	LEPER	SANCTUARY
ASPHYXIATED	DORMANT	MALADJUSTED	SENSIBILITIES
BANISHED	DRONE	MOMENTUM	SUBJECTIVITY
BIGOTED	DYNAMICS	MUSE	SUBMISSION
BURROW	ERRANT	PRUNING	TENACIOUS
COAXES	FOSTER	PSEUDO	VAGUELY
CONUNDRUM	GENETICS	RECESSIVE	VESPIARY
DELINQUENCY	HAZING	REFURBISHED	WAN
DEMERIT	IMBECILES	RELUCTANCE	WISTFUL
DEMURE	INCITING	REPUTATION	
DENSE	INCRIMINATE	RETREAT	

VOCABULARY WORD SEARCH ANSWER KEY - Speak

ALLEGEDLY	DEVIOUS	LEPER	SANCTUARY
ASPHYXIATED	DORMANT	MALADJUSTED	SENSIBILITIES
BANISHED	DRONE	MOMENTUM	SUBJECTIVITY
BIGOTED	DYNAMICS	MUSE	SUBMISSION
BURROW	ERRANT	PRUNING	TENACIOUS
COAXES	FOSTER	PSEUDO	VAGUELY
CONUNDRUM	GENETICS	RECESSIVE	VESPIARY
DELINQUENCY	HAZING	REFURBISHED	WAN
DEMERIT	IMBECILES	RELUCTANCE	WISTFUL
DEMURE	INCITING	REPUTATION	
DENSE	INCRIMINATE	RETREAT	

VOCABULARY CROSSWORD - Speak

Across
2. Stupid or silly people
5. Fake; false; pretend
8. Remote-controlled mechanism
9. Science of heredity and genes
11. Pale in color; sickly-looking
14. Shy; modest; coy
16. Dull or slow-witted
18. Outcast
19. Hole or hideout animals use to take shelter
20. Mark against someone for misconduct
21. Not in sync with one's circumstances
22. Persistent; stubborn; won't give up

Down
1. Nest of social wasps
3. Something or someone that is inspiring to an artist
4. Straying from the right course
5. Cutting; clipping
6. Humiliating; disgracing
7. Make someone appear guilty of a crime
10. Sacred place offering refuge or safety
11. Thoughtful in a sad way; longing
12. Something that nourishes or cares for
13. Deceitful; not straightforward
15. Force or speed of movement
17. Withdraw; go back

VOCABULARY CROSSWORD ANSWER KEY - Speak

			¹V			²I	³M	B	E	C	I	L	⁴E	S		
		⁵P	S	E	⁶U	D	O						R		⁷I	
		R			S		U					⁸D	R	O	N	E
		U			P		⁹G	E	N	E	T	I	¹⁰S		C	
¹¹W	A	N			I		R					A			R		¹²F
I		I			R		A			¹³D		N			I		O
S		N		¹⁴D	E	¹⁵M	U	R	E			C			M		S
T		G		I		O			V			T			I		T
F				N		M			I			U		¹⁶D	E	N	S	E
U		¹⁷R		G		E			O			A			A		R
¹⁸L	E	P	E	R		N		¹⁹B	U	R	R	O	W		T	
		T				T			S			Y			E	
		R				U										
²⁰D	E	M	E	R	I	T		²¹M	A	L	A	D	J	U	S	T	E	D
		A														
		²²T	E	N	A	C	I	O	U	S						

Across
2. Stupid or silly people
5. Fake; false; pretend
8. Remote-controlled mechanism
9. Science of heredity and genes
11. Pale in color; sickly-looking
14. Shy; modest; coy
16. Dull or slow-witted
18. Outcast
19. Hole or hideout animals use to take shelter
20. Mark against someone for misconduct
21. Not in sync with one's circumstances
22. Persistent; stubborn; won't give up

Down
1. Nest of social wasps
3. Something or someone that is inspiring to an artist
4. Straying from the right course
5. Cutting; clipping
6. Humiliating; disgracing
7. Make someone appear guilty of a crime
10. Sacred place offering refuge or safety
11. Thoughtful in a sad way; longing
12. Something that nourishes or cares for
13. Deceitful; not straightforward
15. Force or speed of movement
17. Withdraw; go back

VOCABULARY MATCHING 1 - Speak

___ 1. REFURBISHED   A. Fake; false; pretend

___ 2. WISTFUL   B. Hole or hideout animals use to take shelter

___ 3. INCONSPICUOUS   C. Not noticeable

___ 4. VESPIARY   D. Thoughtful in a sad way; longing

___ 5. FLOUNDERING   E. Cutting; clipping

___ 6. PSEUDO   F. Mark against someone for misconduct

___ 7. GENETICS   G. Intolerant of any other beliefs or opinions

___ 8. DEMERIT   H. Sacred place offering refuge or safety

___ 9. MALADJUSTED   I. Something that nourishes or cares for

___ 10. SUBMISSION   J. Remote-controlled mechanism

___ 11. RECESSIVE   K. Social, intellectual, or physical forces that characterize a system or group

___ 12. INCITING   L. Something or someone that is inspiring to an artist

___ 13. BIGOTED   M. Nest of social wasps

___ 14. REVOLUTIONARY   N. Failure to fulfill a duty or obligation; something wrongful or illegal

___ 15. DYNAMICS   O. Going to the back; a gene that does not produce

___ 16. BURROW   P. Acting clumsily or in confusion

___ 17. DELINQUENCY   Q. Stirring up (trouble); egging-on

___ 18. DRONE   R. Unwillingness; resistance

___ 19. SANCTUARY   S. How the public views or regards an individual

___ 20. REPUTATION   T. Persistent; stubborn; won't give up

___ 21. RELUCTANCE   U. Supporting radical change or innovation

___ 22. MUSE   V. Science of heredity and genes

___ 23. PRUNING   W. Surrendering power to another

___ 24. FOSTER   X. Made clean, bright, or fresh again

___ 25. TENACIOUS   Y. Not in sync with one's circumstances

VOCABULARY MATCHING 1 ANSWER KEY - Speak

X - 1. REFURBISHED
D - 2. WISTFUL
C - 3. INCONSPICUOUS
M - 4. VESPIARY
P - 5. FLOUNDERING
A - 6. PSEUDO
V - 7. GENETICS
F - 8. DEMERIT
Y - 9. MALADJUSTED
W - 10. SUBMISSION
O - 11. RECESSIVE
Q - 12. INCITING
G - 13. BIGOTED
U - 14. REVOLUTIONARY
K - 15. DYNAMICS
B - 16. BURROW
N - 17. DELINQUENCY
J - 18. DRONE
H - 19. SANCTUARY
S - 20. REPUTATION
R - 21. RELUCTANCE
L - 22. MUSE
E - 23. PRUNING
I - 24. FOSTER
T - 25. TENACIOUS

A. Fake; false; pretend
B. Hole or hideout animals use to take shelter
C. Not noticeable
D. Thoughtful in a sad way; longing
E. Cutting; clipping
F. Mark against someone for misconduct
G. Intolerant of any other beliefs or opinions
H. Sacred place offering refuge or safety
I. Something that nourishes or cares for
J. Remote-controlled mechanism
K. Social, intellectual, or physical forces that characterize a system or group
L. Something or someone that is inspiring to an artist
M. Nest of social wasps
N. Failure to fulfill a duty or obligation; something wrongful or illegal
O. Going to the back; a gene that does not produce
P. Acting clumsily or in confusion
Q. Stirring up (trouble); egging-on
R. Unwillingness; resistance
S. How the public views or regards an individual
T. Persistent; stubborn; won't give up
U. Supporting radical change or innovation
V. Science of heredity and genes
W. Surrendering power to another
X. Made clean, bright, or fresh again
Y. Not in sync with one's circumstances

VOCABULARY MATCHING 2 - Speak

___ 1. ASPHYXIATED A. Failure to fulfill a duty or obligation; something wrongful or illegal
___ 2. DEGRADING B. Science of heredity and genes
___ 3. WISTFUL C. Unwillingness; resistance
___ 4. SUBJECTIVITY D. Suffocated; smothered; choked
___ 5. ERRANT E. Thoughtful in a sad way; longing
___ 6. SANCTUARY F. Withdraw; go back
___ 7. DELINQUENCY G. Remote-controlled mechanism
___ 8. SUBMISSION H. Social, intellectual, or physical forces that characterize a system or group
___ 9. FLOUNDERING I. Persistent; stubborn; won't give up
___ 10. DYNAMICS J. Abusing newcomers with humiliating tricks and ridicule before they become a part of the group
___ 11. VAGUELY K. Intolerant of any other beliefs or opinions
___ 12. INCONSPICUOUS L. Surrendering power to another
___ 13. RETREAT M. Sacred place offering refuge or safety
___ 14. MOMENTUM N. Force or speed of movement
___ 15. PSEUDO O. Pale in color; sickly-looking
___ 16. BIGOTED P. Not clear or definite
___ 17. HAZING Q. Based on personal feelings rather than facts
___ 18. DRONE R. Fake; false; pretend
___ 19. GENETICS S. Inactive; asleep
___ 20. RELUCTANCE T. Keeping the same behavior, form, pattern, or principles
___ 21. WAN U. Mark against someone for misconduct
___ 22. CONSISTENCY V. Humiliating; disgracing
___ 23. TENACIOUS W. Straying from the right course
___ 24. DORMANT X. Acting clumsily or in confusion
___ 25. DEMERIT Y. Not noticeable

VOCABULARY MATCHING 2 ANSWER KEY - Speak

D - 1. ASPHYXIATED   A. Failure to fulfill a duty or obligation; something wrongful or illegal

V - 2. DEGRADING   B. Science of heredity and genes

E - 3. WISTFUL   C. Unwillingness; resistance

Q - 4. SUBJECTIVITY   D. Suffocated; smothered; choked

W - 5. ERRANT   E. Thoughtful in a sad way; longing

M - 6. SANCTUARY   F. Withdraw; go back

A - 7. DELINQUENCY   G. Remote-controlled mechanism

L - 8. SUBMISSION   H. Social, intellectual, or physical forces that characterize a system or group

X - 9. FLOUNDERING   I. Persistent; stubborn; won't give up

H - 10. DYNAMICS   J. Abusing newcomers with humiliating tricks and ridicule before they become a part of the group

P - 11. VAGUELY   K. Intolerant of any other beliefs or opinions

Y - 12. INCONSPICUOUS   L. Surrendering power to another

F - 13. RETREAT   M. Sacred place offering refuge or safety

N - 14. MOMENTUM   N. Force or speed of movement

R - 15. PSEUDO   O. Pale in color; sickly-looking

K - 16. BIGOTED   P. Not clear or definite

J - 17. HAZING   Q. Based on personal feelings rather than facts

G - 18. DRONE   R. Fake; false; pretend

B - 19. GENETICS   S. Inactive; asleep

C - 20. RELUCTANCE   T. Keeping the same behavior, form, pattern, or principles

O - 21. WAN   U. Mark against someone for misconduct

T - 22. CONSISTENCY   V. Humiliating; disgracing

I - 23. TENACIOUS   W. Straying from the right course

S - 24. DORMANT   X. Acting clumsily or in confusion

U - 25. DEMERIT   Y. Not noticeable

VOCABULARY JUGGLE LETTERS 1 - Speak

1. IVAUNLORYOTRE = 1. _____
   Supporting radical change or innovation

2. EMSU = 2. _____
   Something or someone that is inspiring to an artist

3. IDMETER = 3. _____
   Mark against someone for misconduct

4. AODRTINNINTCIO = 4. _____
   Teaching someone to accept an idea or principle without criticism

5. CRTAUANSY = 5. _____
   Sacred place offering refuge or safety

6. OMUEMMNT = 6. _____
   Force or speed of movement

7. BFIEERDSUHR = 7. _____
   Made clean, bright, or fresh again

8. RTARNE = 8. _____
   Straying from the right course

9. UMDORNCUN = 9. _____
   Difficult problem; dilemma

10. TPIDESXYAHA =10. _____
    Suffocated; smothered; choked

11. SECCNEONCI =11. _____
    Inner sense of what is right or wrong

12. LSTIFWU =12. _____
    Thoughtful in a sad way; longing

13. GGRDAINED =13. _____
    Humiliating; disgracing

14. NTENMIAICRI =14. _____
    Make someone appear guilty of a crime

15. RELNGOUDFIN =15. _____
    Acting clumsily or in confusion

VOCABULARY JUGGLE LETTERS 1 ANSWER KEY - Speak

1. IVAUNLORYOTRE = 1. REVOLUTIONARY
   Supporting radical change or innovation

2. EMSU = 2. MUSE
   Something or someone that is inspiring to an artist

3. IDMETER = 3. DEMERIT
   Mark against someone for misconduct

4. AODRTINNINTCIO = 4. INDOCTRINATION
   Teaching someone to accept an idea or principle without criticism

5. CRTAUANSY = 5. SANCTUARY
   Sacred place offering refuge or safety

6. OMUEMMNT = 6. MOMENTUM
   Force or speed of movement

7. BFIEERDSUHR = 7. REFURBISHED
   Made clean, bright, or fresh again

8. RTARNE = 8. ERRANT
   Straying from the right course

9. UMDORNCUN = 9. CONUNDRUM
   Difficult problem; dilemma

10. TPIDESXYAHA =10. ASPHYXIATED
    Suffocated; smothered; choked

11. SECCNEONCI =11. CONSCIENCE
    Inner sense of what is right or wrong

12. LSTIFWU =12. WISTFUL
    Thoughtful in a sad way; longing

13. GGRDAINED =13. DEGRADING
    Humiliating; disgracing

14. NTENMIAICRI =14. INCRIMINATE
    Make someone appear guilty of a crime

15. RELNGOUDFIN =15. FLOUNDERING
    Acting clumsily or in confusion

VOCABULARY JUGGLE LETTERS 2 - Speak

1. CYSDAIMN = 1. _____
Social, intellectual, or physical forces that characterize a system or group

2. NTARMDO = 2. _____
Inactive; asleep

3. DIEVUSO = 3. _____
Deceitful; not straightforward

4. IRGPUNN = 4. _____
Cutting; clipping

5. ESEDN = 5. _____
Dull or slow-witted

6. UEERDM = 6. _____
Shy; modest; coy

7. VSSEIRCEE = 7. _____
Going to the back; a gene that does not produce

8. SYCNNESOITC = 8. _____
Keeping the same behavior, form, pattern, or principles

9. PYVIERSA = 9. _____
Nest of social wasps

10. MNCDORNUU = 10. _____
Difficult problem; dilemma

11. UPDSOE = 11. _____
Fake; false; pretend

12. ITSFUWL = 12. _____
Thoughtful in a sad way; longing

13. RDEON = 13. _____
Remote-controlled mechanism

14. TTREUNOIAP = 14. _____
How the public views or regards an individual

15. NAW = 15. _____
Pale in color; sickly-looking

VOCABULARY JUGGLE LETTERS 2 ANSWER KEY - Speak

1. CYSDAIMN = 1. DYNAMICS
   Social, intellectual, or physical forces that characterize a system or group

2. NTARMDO = 2. DORMANT
   Inactive; asleep

3. DIEVUSO = 3. DEVIOUS
   Deceitful; not straightforward

4. IRGPUNN = 4. PRUNING
   Cutting; clipping

5. ESEDN = 5. DENSE
   Dull or slow-witted

6. UEERDM = 6. DEMURE
   Shy; modest; coy

7. VSSEIRCEE = 7. RECESSIVE
   Going to the back; a gene that does not produce

8. SYCNNESOITC = 8. CONSISTENCY
   Keeping the same behavior, form, pattern, or principles

9. PYVIERSA = 9. VESPIARY
   Nest of social wasps

10. MNCDORNUU = 10. CONUNDRUM
    Difficult problem; dilemma

11. UPDSOE = 11. PSEUDO
    Fake; false; pretend

12. ITSFUWL = 12. WISTFUL
    Thoughtful in a sad way; longing

13. RDEON = 13. DRONE
    Remote-controlled mechanism

14. TTREUNOIAP = 14. REPUTATION
    How the public views or regards an individual

15. NAW = 15. WAN
    Pale in color; sickly-looking

www.ingramcontent.com/pod-product-compliance
Lightning Source LLC
Chambersburg PA
CBHW051407070526
44584CB00023B/3331